Photorefractive Photonics 2017

Journal of Physics: Conference Series Volume 867

Qingdao, China
17 - 20 July 2017

Editors:

Guoquan Zhang
Weijin Kong
Jingjun Xu

ISBN: 978-1-5108-4422-3
ISSN: 1742-6588

Printed from e-media with permission by:

Curran Associates, Inc.
57 Morehouse Lane
Red Hook, NY 12571

Some format issues inherent in the e-media version may also appear in this print version.

Printed by Curran Associates, Inc. (2017)

For permission requests, please contact the Institute of Physics at the address below.

Institute of Physics
Dirac House, Temple Back
Bristol BS1 6BE UK

Phone: 44 1 17 929 7481
Fax: 44 1 17 920 0979

techtracking@iop.org

Additional copies of this publication are available from:

Curran Associates, Inc.
57 Morehouse Lane
Red Hook, NY 12571 USA
Phone: 845-758-0400
Fax: 845-758-2633
Email: curran@proceedings.com
Web: www.proceedings.com

Table of contents

Volume 867

Photorefractive Photonics 2017
17–20 July 2017, Qingdao, Shandong, China

Accepted papers received: 26 June 2017
Published online: 5 July 2017

Preface

Papers

IOP Conf. Series: Journal of Physics: Conf. Series **867** (2017) 011001 doi:10.1088/1742-6596/867/1/011001

Proceeding of Photorefractive Photonics 2017

Preface

This volume of *Journal of Physics: Conference Series* represents a selection of the contributions presented in the Photorefractive Photonics 2017 (PR17) conference which was held in Qingdao, Shandong, China from July 17 through July 20, 2017. The conference is the successor of the traditional Photorefractive (PR) Conference Series: the International Conference on Photorefractive Effects, Materials and Devices. The first PR Conference was held in Los Angeles, California, USA in 1987. After that, the PR Conference was held every two years around the world: Aussois, France (1989), Beverly, Massachusetts, USA (1991), Kiev, Ukraine (1993), Aspen Lodge, Colorado, USA (1995), Chiba, Japan (1997), Elsinore, Denmark (1999), Delevan, Wisconsin, USA (2001), Nice, France (2003), Sanya, Hainan, China (2005), Olympic Valley, California, USA (2007), Bad Honnef, Germany (2009), Ensenada, Mexico (2011), Winchester, UK (2013) and Villars, Switzerland (2015). The Conference was renamed to be Photorefractive Photonics Conference in 2015. The PR17 conference was organized by Nankai University, Qingdao University and Qingdao Society of Physics.

The scientific topics of the PR17 conference include, but not limited to, wave mixing and its applications (T1), nonlinear light-matter interaction and its applications (T2), material engineering and photonic micro-structures (T3), photorefractive lattice photonics (T4), digital photorefractive photonics (T5), and novel photorefractive effects, materials and applications (T6). Note that both traditional topics such as photorefractive wave mixing and new topics such as energy photonics and nano/micro-structures based on photorefractive effects and materials are included, especially, basic issues such as the influence of pyroelectric effect on the photorefractive wave-mixing are considered, and much attention was paid to the development of new applications of photorefractive effects and materials, which reflect the current state of the art of photorefractive research.

We would like to thank all scientific topic chairs for their suggestions of invited talks, Prof. Fang Bo for his patience to communicate with the conference attendees, and Dr. Feng Xia for his technical support during the conference preparation. We are also very indebted to the international scientific committees, the conference scientific topic chairs and the local organizing committee for the success of this conference and the quality of this volume.

This PR17 conference is financially supported by the National Natural Science Foundation of China, Nankai University and Qingdao University.

Editors
Guoquan Zhang, Nankai University
Weijin Kong, Qingdao University
Jingjun Xu, Nankai University

IOP Conf. Series: Journal of Physics: Conf. Series **867** (2017) 011002 doi:10.1088/1742-6596/867/1/011002

Peer review statement

All papers published in this volume of *Journal of Physics: Conference Series* have been peer reviewed through processes administered by the proceedings Editors. Reviews were conducted by expert referees to the professional and scientific standards expected of a proceedings journal published by IOP Publishing.

IOP Conf. Series: Journal of Physics: Conf. Series **867** (2017) 012001 doi:10.1088/1742-6596/867/1/012001

Impact of piezoelectric and flexoelectric effects on the wave mixing in photosensitive crystals

S M Shandarov[1] and V V Shepelevich[2]

[1]Tomsk State University of Control Systems and Radioelectronics, 40 Lenin Ave., Tomsk 634050, Russia
[2]I.P. Shamyakin Mozyr State Pedagogical University, 28 Studencheskaya Str., Mozyr 247760, Belarus

E-mail: stanislavshandarov@gmail.com

Abstract. We consider the main phenomena relating to elastic fields accompanying through the converse piezoelectric and flexoelectric effects the electric ones created in photosensitive crystals by inhomogeneous distributions of light intensity as a results of dynamic hologram formation.

1. Introduction

The non-uniform electric fields of photorefractive holograms may be accompanied by the elastic fields owing to converse piezoelectric and flexoelectric effects (see, e.g., [1] and [2], and references therein). In this report we describe the main phenomena connected with an occurrence of such elastic fields both in the boundless crystals with no center of symmetry and in the semi-bounded ones.

2. General equations

The elastic fields induced in the crystal by the space-charge field of photorefractive grating should satisfy the elastostatic equation [3]

$$\frac{\partial}{\partial x_j} T_{ij} = 0, \tag{1}$$

where T_{ij} are the tensor components of elastic stress. In the absence of spatial dispersion the relevant equations of state determining reciprocal relationship between quasi-static electric and elastic fields in crystals with no center of symmetry in the general case can be represented from equations taking into account the nonlocality of response [4] as

$$T_{ij} = C^{E}_{ijkl} S_{kl} - e_{mij} E_m + f_{ijmr} \frac{\partial E_m}{\partial x_r}, \tag{2}$$

$$D_n = e_{nkl} S_{kl} + \varepsilon^{S}_{nm} E_m + f_{klnr} \frac{\partial S_{kl}}{\partial x_r}, \tag{3}$$

where S_{kl} are the tensor components of elastic strain, E_m and D_n are the vector components of the electric field and electric displacement, and C^E_{ijkl}, ε^S_{nm}, and e_{mij} are the tensor components of the elastic modulus, the static dielectric permittivity and the piezoelectric constant of the crystal, respectively. The last gradient terms in Eqs. (2) and (3) define respectively the converse and direct flexoelectriceffects describable by fourth-rank flexoelectric tensor with components f_{ijmr}.

The solution for elastic fields of a photorefractive grating formed by continuous light pattern in boundless crystals can be obtained, as in [3] for converse piezoelectric effect solely, from the equation of elastostatics and Eq. (1). It is significant that an elastic strain distribution determined by converse piezoelectric effect is in-phase with electric field [3], whereas the converse flexoelectricity produces the similar distribution, but shifted by $\pi/2$. The magnitude of nonshifted component of the elastic strain is independent on fringe spacing Λ of the grating and is determined by anisotropy of piezoelectric properties of the crystal [1, 3]. By contrast, the magnitude of flexoelectrically induced elastic fields varies inversely with Λ and can be different from zero in centrosymmetrical photosensitive media also [2].

As can be seen from Eq. (2), the elastic strains and the gradient of ones provide the additional contribution to the electric polarization of crystal, which depends on the crystal anisotropy and on the parameters of photorefractive grating. To describe the formation of space-charge field of photorefractive grating with taking into account the piezoelectric contribution, the effective static dielectric permittivity ε' can be used [3]. This renormalized permittivity satisfies the inequality $\varepsilon^S \le \varepsilon' \le \varepsilon^T$, where ε^S and ε^T are the effective static dielectric permittivity without taking into account the piezoelectric effect for clamped and unclamped crystals, respectively.

The contribution of elastic fields under consideration to the perturbation of the light-frequency dielectric tensor of a crystal, which is additional to the conventional electro-optic one, is determined by elasto-optic effect [1, 2]. The nonshifted piezoelectric component of elasto-optic contribution, which was well-known formerly (see, e.g., [1] and [5]), may even overtop the electro-optic one for certain directions of the grating vector in ferroelectric photorefractive crystals such as $BaTiO_3$.

In the case of the electric and elastic fields of photorefractive gratings in semi-bounded piezoelectric crystals, the relevant boundary conditions must be taking into account [6].

3. Conclusions

We have considered the main phenomena related to the elastic fields accompanying the electric ones created by inhomogeneous distributions of light intensity in photosensitive and photorefractive crystals owing to the converse piezoelectric and flexoelectric effects.

Acknowledgments

This work was supported by the Ministry of the Education of Belarus (project 1.2.01 of the State Program “Photonics and Opto- and Microelectronics), the State Assignment of the Ministry of Education and Science of Russian Federation for 2017 and the RFBR (grant No. 16-29-14046-ofi_m).

References

[1] Shepelevich V V 2011 *J. Appl. Spectroscopy* **78** 461
[2] Shandarov S M, Smakov S S, Zuev P V, Burimov N I, Kargin Yu F, Shepelevich V V, Ropot P I, Gudelev V G 2013 *J. Opt. Technol.* **80** 409
[3] Shandarov S 1992 *Appl. Phys. A* **55** 91
[4] Vuzhva A D, Lyamov V E 1977 *Sov. Phys. Crystallogr.* **22** 73
[5] Zgonik M, Nakagava K, Günter P 1995 *J. Opt. Soc. Am. B* **95** 699
[6] Burimov N I, Shandarov S M 2006 *Phys. Solid State* **48** 524

Time resolved nonlinear response of $Sn_2P_2S_6$:Sb to nanosecond pulse excitation

A Shumelyuk[1], Y Skrypka[1], S Odoulov[1],
S A Basun[2,3], L E Halliburton[4] and D R Evans[2]

[1] Institute of Physics, Science Ave. 46, 03650 Kyiv, Ukraine
[2] Air Force Research Laboratory, Wright-Patterson AFB, OH 45443, USA
[3] Azimuth Corp., 4027 Colonel Glenn Hwy., Beavercreek, Ohio 45431, USA
[4] Department of Physics, West Virginia University, Morgantown, WV 26506, USA

Corresponding author's e-mail address: odoulov@iop.kiev.ua

Abstract. The temporal variations of SPS:Sb transmission are measured within a 10 orders of magnitude time range (from a few nanoseconds to ten seconds) with a red diode laser beam, after illumination of the sample with a nanosecond pulse of green light. The measurements revealed a sequence of darkening and bleaching processes that are sensitive to the intensities of pre-illumination and testing beams, or the sample temperature, or both. This allowed for identifying particular nonlinear processes that manifest themselves in changes of transmission.

1. Introduction

$Sn_2P_2S_6$:Sb (SPS:Sb) is a fast and efficient photorefractive crystal sensitive in red up to 800 nm [1]. Its response can be enhanced by pre-illumination to white light [2]; its gain depends on the recording intensity. The comparison of photorefractive and EPR data allowed for identifying Sb as a secondary photorefractive center responsible for optical sensitizing [3]. In the present contribution, we study the variation of SPS:Sb transmission at 633 nm within the interval from 10^{-9} s to 10 s, after its illumination with 20 ns pulse of green light. Such measurements are done for different intensities of the exciting and testing radiation, within the temperature range from -109 to 40°C.

2. Experiment

A 2.4 mm-thick sample of SPS that contains 1% of antimony was fixed on a copper cold finger in a vacuum chamber; the temperature control was ensured with a cold nitrogen flow and a heater. The chamber had two optical windows to illuminate the sample with a short powerful pulse from Continuum Panther OPO (532 nm, 1.3 J/cm^2, 20 ns) and to monitor the sample transmission with a diode laser beam (0.633 nm, 1.5 W/cm^2). Two polarizers and two half-wave phase retarders allowed for polarization adjustments of both beams and for the control of their intensities.

Figure 1 shows a representative example of temporal variations of the diode laser red beam transmitted through SPS:Sb sample at ambient temperature during and after exposure to the green pulse, plotted in a log time scale. Both, laser beams are linearly polarized along the sample x-axis. An entire temporal range was measured in several attempts with different sampling times and different total times of measurement, to ensure a sufficient temporal resolution in every particular time interval.

During the pulse duration (≈ 20 ns) the sample transmission is gradually decreasing, pointing to photoexcitation of free carriers that might further form polarons [4].

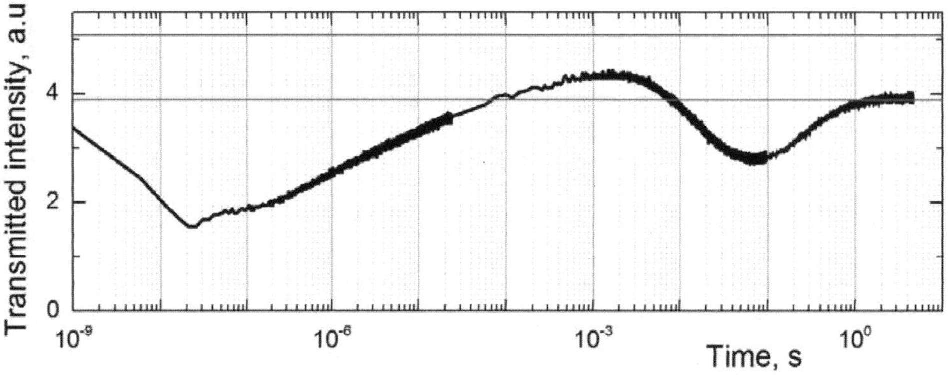

Figure 1. Temporal dynamics of SPS:Sb sample transmission at 633 nm and ambient temperature during and after 20 ns green pulse illumination (black curve). Two red lines mark the initial level of red beam transmission (upper line) and transmission level of partially exhausted for photorefractive scattering red beam in the virgin sample (with no green pulse exposure).

These photoinduced entities decay within the time range from 0.1 ms to 1 ms or even longer. The exact time is shadowed by a new extinction process that starts from a few milliseconds and is related to the development of photorefractive scattering known as beam fanning. Finally, we observe the second bleaching process from ≈ 50 ms to several seconds, with a contribution of the formation of the compensation gratings by moving charges of opposite sign [5].

The next experiments aimed to extract the characteristic build-up and decay times for three processes that occur after termination of the exciting green pulse. The sets of measurements were conducted for every process at different red light intensities and at different sample temperatures. This allowed for identification of their physical origin and for estimating the activation energies from the Arrhenius plots. The comparison of the obtained data with the activation energies extracted from EPR measurements [3,4] allowed for the conclusion that the initial light-induced absorption is caused by hole polarons (Sn^{2+} with a trapped hole [4]), while it is the antimony dopant which is responsible, indirectly and directly, for beam fanning and the development of compensation gratings [5].

The question may arise, why in the present experiment we don't see 14 s decay time as observed in [2] with 633 nm pre-exposure. This might be due to two reasons. In a few seconds the cw red beam may generate a comparable Sb^{2+} density as that done by the green-pulse pre-exposure so that the Sb^{2+} density is close to its steady-state value. The second reason might be related to the hyperbolic decay character of the $Sb^{2+} \rightarrow Sb^{3+}$ relaxation revealed in [6] at low temperatures.

References

[1] Bach T, Jazbinsek M, Montemezzani G, Gunter P, Grabar A and Vysochanskii Yu 2007 *J. Opt. Soc. Am. B* **24** 1535

[2] Evans D R, Shuymelyuk A, Cook G and Odoulov S 2011 *Opt. Lett.* **36** 454

[3] Brant A T, Halliburton L E, Basun S A, Grabar A A, Odoulov S G, Shumelyuk A, Giles N C and Evans D R 2012 *Phys. Rev. B* **86** 134109

[4] Brant A T, Halliburton L E, Giles N C, Basun S A, Grabar A A and Evans D R 2013 *J. Phys.: Condens. Matter* **25**, 205501

[5] Skrypka Y, Shumelyuk A, Odoulov S, Basun S and Evans D R 2017 *Opt. Mat. Express*, **7**, 1414

[6] Basun S A, Halliburton L E and Evans D R 2017 2017 *Appl. Phys. Lett.* **110,** 052903

IOP Conf. Series: Journal of Physics: Conf. Series **867** (2017) 012003 doi:10.1088/1742-6596/867/1/012003

Dynamic grating mediated energy transfer between two sub-picosecond pulses in $LiNbO_3$

S Nolte[1], B Bourdon[1], F Freytag[1], M Imlau[1], A Shumelyuk[2] and S Odoulov[2]

[1] Department of Physics, Osnabrueck University,
Barbarastrasse 7, D-49076 Osnabrueck, Germany
[2] Institute of Physics, Science Ave. 46, 03650 Kyiv, Ukraine

Corresponding author's e-mail address: stenolte@uni-osnabrueck.de

Abstract. Two ultrashort pulses that interact in a nonlinear medium can excite a grating of the nonlinear absorption and refraction and their characteristics can be modified as a result of selfdiffraction from this grating. In this work, we adress an energy transfer between pulses of different energies and/or different frequencies within the visible spectral range in congruent, nominally undoped $LiNbO_3$ aiming to reveal new features that distinguish this process from the known transient and steady-state coupling of cw laser beams.

1. Introduction

In this presentation we describe a study of the direct energy exchange between ultrashort femtosecond pulses in $LiNbO_3$, the crystal which possesses multiple quite pronounced optical nonlinearities. Apart from the instanteneous high-frequency Kerr nonlinearity and two-photon absorbtion nonlinearity [1], several inertial nonlinear responses have been revealed including those which are due to photoexcitation of free carriers and formation of various kinds of polarons [2].This crystal was the first in which a photorefractive nonlinearity and grating-assisted intensity redistribution of the recording waves has been discovered, see, e.g., [3]. We undertake this study to estimate how strong the coupling of subpicosecond pulses can be in $LiNbO_3$ and to establish the underlying wave mixing process(es) that is(are) responsible for energy redistribution between the interacting pulses.

2. Experimental Setup

The femtosecond laser system consists of a Ti:Sapphire oscillator with amplifier as a pump source for two independent tunable Optical Parametric Amplifiers (OPAs) [cf. [1]]. For the study of frequency degenerate interaction, the output of a single OPA was divided by a beamsplitter, while the use of two OPAs allowed for introducing any desirable frequency detuning between the recording pulses.

In the experiment, the two light pulses impinge upon the sample in a plane normal to the axis of spontaneous polarization and are polarized along this axis to avoid possible contribution of space-charge (photorefractive) gratings in investigated coupling processes. The angle of the beam intersection in air was chosen to be less than 5° to ensure a good spatial overlap of two 100 fs-long pulses. With the help of an optical delay line in one of two intersecting light beams the recording pulses could be perfectly well adjusted in time or deliberately mismatched with a controllable time delay. A double grating setup was used as a compressor/stretcher to change the chirp. This pulse chirp opens the possibility to introduce a frequency difference by changing the temporal mismatch, even in the case of the recording pulses with identical spectra.

3. Results & Discussion

Figure 1a shows the relative sample transmission T for different pulse delays Δt for two interacting pulses with identical pulse durations, pulse energies, chirp coefficients and spectra (centred at 590 nm). The upper and lower graphs differ in the sign of chirp (and pulse energy). In both graphs, pronounced energy redistribution (± 20 %) is observed at Δt close to FWHM of the pulses themselves. Within the time interval where pulses interact a red-shifted portion of one pulse is always amplified in expense of the blue-shifted portion of the other pulse. Thus, for a fixed Δt the direction of energy flow is inverted if the sign of the chirp becomes opposite. This type of pulse energy coupling via diffraction from a moving grating is already well known (see, e.g., for plasma gratings in gases [5]).

To confirm that it is just a frequency difference that governs this particular energy coupling process we used two pulses with 5 nm difference in wavelength. Instead of two, well separated in Δt energy coupling areas (Figure 1a) we obtained only one high coupling area centered at $\Delta t = 0$.

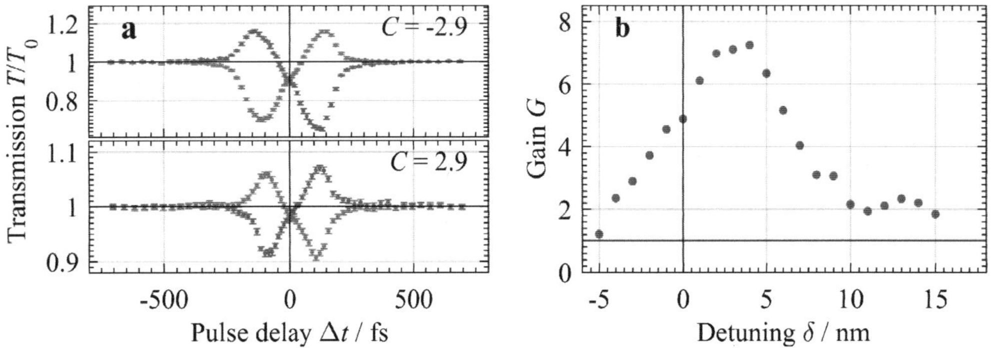

Figure 1. (a) Pulse delay dependence of transmission $T(\Delta t)$ for two identical interacting pulses (see text) with pulse durations $\tau = 200$ fs (FWHM) and opposite chirp parameters C. (b) Wavelength detuning $\delta = \lambda_{probe} - \lambda_{pump}$ dependence of the gain G for a weak probe pulse.

The next set of experiments consisted in gain measurement of a low signal probe pulse in presence of a strong pump pulse with tunable frequency. Figure 1b shows that nonzero coupling gain of a weak probe exists also in case of strict degeneracy, when both, the pump and probe pulse have identical spectra centered at 488 nm. This confirms the presence of another coupling process that is sensitive not to the detuning in pulse frequencies but to pulse difference in intensity. We believe this is a manifestation of a transient energy transfer (TET) [5] known for coupling of long pulses in media with inertial nonlinearities but never considered before for femtosecond pulses. Thus the detuning dependence of Figure 1b consists of two contributions, a δ-symmetric (odd function) dependence from TET and δ-anti-symmetric (even function) contribution from coupling of frequency nondegenerate pulses.

The physical processes that are causing these optical nonlinearities are not yet established, but we know that for observing TET the decay time of the nonlinearity should be longer comparing to the pulse duration. However, for efficient coupling of frequency nondegenerate pulses, on the opposite, this time should be comparable or shorter than the pulse duration.

Acknowledgements The financial support of the Deutsche Forschungs Gemeinschaft (DFG) via projects IM37/5-2, INST 190/137-1 FUGG and INST 190/165-1 FUGG is gratefully acknowledged.

References

[1] Badorreck H, Nolte S, Freytag F, Bäune P, Dieckmann V and Imlau M 2015 Opt. Mat. Express **5**, 2729
[2] Imlau M, Badorreck H and Merschjann C 2015 Appl. Phys. Rev. **2**, 040606
[3] *Photorefractive materials and their applications*, vol. 1, 2, and 3, Günter P and Huignard J-P (editors), Springer series in Optical Science, Springer, 2007.

IOP Conf. Series: Journal of Physics: Conf. Series **867** (2017) 012003 doi:10.1088/1742-6596/867/1/012003

[4] Liu Y, Durand M, Chen S, Houdard A, Prade B, Forrestier B and Mysyrowicz A 2010 Phys. Rev. Lett. **105**, 055003
[5] Kukhtarev N, Markov V and Odoulov S 1977 Opt. Commun. **23** 338

IOP Conf. Series: Journal of Physics: Conf. Series **867** (2017) 012004 doi:10.1088/1742-6596/867/1/012004

Accelerating space-charge gratings in wide-bandgap semiconductors

Igor Sokolov, Mikhail Bryushinin

Ioffe Institute, Politekhnicheskaya 26, 194021 St. Petersburg, Russia

i.a.sokolov@mail.ioffe.ru

Abstract. The excitation of the non-steady-state photoelectromotive force and two-wave mixing signals is theoretically investigated for the case of uniformly accelerated motion of the recording light pattern. Such illumination is usually created by the linear frequency modulation of the interfering light beams. For the both effects we predict the pulsed response arising at the moments, when the velocities of the interference pattern and space-charge wave coincide. The application of the effects in laser Doppler velocimeters and accelerometers is discussed.

1. Introduction

The interference pattern in the two-wave mixing (TWM) and non-steady-state photoelectromotive force (photo-EMF) experiments is usually created by two coherent light beams, which can be frequency shifted or phase modulated by a sinusoidal signal. The constant frequency shift can arise because of the Doppler effect and sinusoidal phase modulation appears in the beam reflected from a vibrating object. In this study we try to extend the variety of the effects observed in photoconductive media introducing linear frequency modulation (LFM) in the light beams forming the interference pattern. We investigate theoretically and experimentally the manifestation of such illumination in the non-steady-state photo-EMF and two-wave mixing effects.

2. Analysis of the effect

The main idea of the experiment is sketched in Fig. 1. The crystal is illuminated by two light waves, the signal beam is frequency-modulated with chirp rate A: $f(t)=At$. These beams produce the accelerating interference pattern with the intensity $I(x,t)=I_0[1+0.5m(t)e^{iKx}+0.5m^*(t)e^{-iKx}]$, where $m(t)=me^{i\varphi(t)}$ and $\varphi(t)=-At^2$.

The analysis is performed using the simplest model of electron semiconductor with one type of partially compensated donor centers [1]. The solution of the balance, continuity and Poisson equations provides the following expressions for the photo-EMF signal:

$$j(t) = -\frac{m^2\sigma_0 E_D}{2(1+K^2L_D^2)}\mathrm{Im}\Xi(t) \qquad \text{for } E_0=0, \tag{1}$$

$$j(t) = -\frac{m^2\sigma_0 E_0}{2}|\Xi(t)|^2 \qquad \text{for } E_0>>E_D,\ KL_0>>1+K^2L_D^2, \tag{2}$$

and for the power of the time-dependent component of the signal beam behind the crystal (TWM signal):

$$P_{S\sim}(t) = m(P_RP_S)^{1/2}\delta dE_D\mathrm{Re}\Xi(t) \qquad \text{for } E_0=0, \tag{3}$$

$$P_{S\sim}(t) = m(P_RP_S)^{1/2}\delta dE_0\mathrm{Im}\Xi(t) \qquad \text{for } E_0>>E_D,\ KL_0>>1+K^2L_D^2 \tag{4}$$

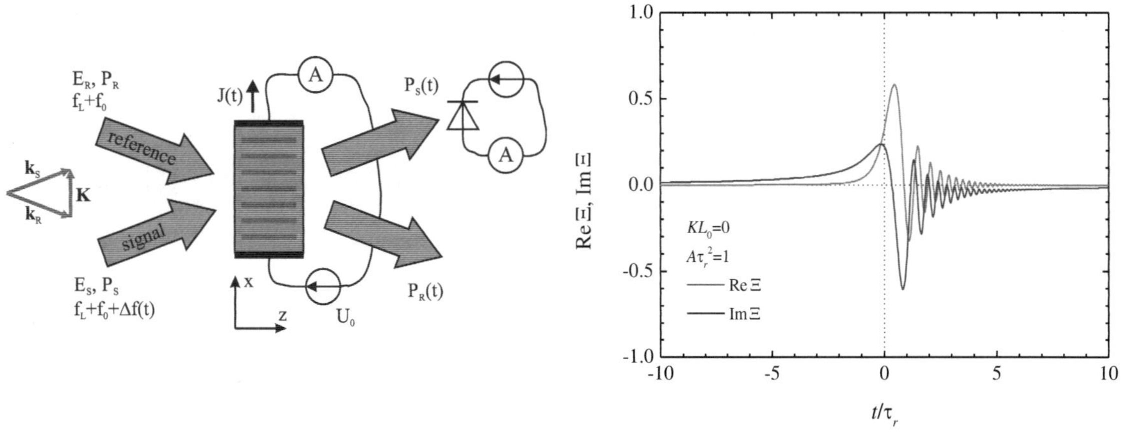

Figure 1. Scheme of the space charge formation and excitation of the non-steady-state photo-EMF and TWM signals by the frequency modulated light (left). Time dependencies of the real and imaginary parts of function $\Xi(t)$ (normalized TWM and photo-EMF signals) calculated for the diffusion regime (right).

where

$$\Xi(t) = \frac{1}{2\sqrt{A}\tau_{sc}^{*}} \exp\left[-\frac{t}{\tau_{sc}^{*}} + i\frac{\pi}{4} + \frac{i}{4\pi A \tau_{sc}^{*2}} - i\pi A t^{2}\right] \mathrm{Erf}\left[-\sqrt{\frac{\pi A}{i}}t + \frac{1}{2\tau_{sc}^{*}}\sqrt{\frac{i}{\pi A}}\right] \quad (5)$$

is the dimensionless function, σ_0 is the average photoconductivity, E_0 is the applied dc electric field, E_D is the diffusion field, $\tau_{sc}=\tau_M(1+K^2L_D^2-iKL_0)$ is the time constant of the space-charge grating, τ_M is the Maxwell relaxation time, L_D and L_0 are the electron diffusion and drift lengths, P_R and P_S are the powers of the reference and signal beams, δ is the constant using in the TWM theory, d is the crystal thickness, and Erf is the complementary error function.

The time evolution of the normalized TWM and photo-EMF signals calculated for the diffusion regime (E_0=0) is shown in Fig. 1. The dependencies are plotted versus dimensionless time t/τ_r, where $\tau_r=\tau_M[(1+K^2L_D^2)^2+K^2L_0^2]/(1+K^2L_D^2)$ is the recording time. The signals appear at the moments, when interference pattern slows down and stops. At low chirp rates the TWM pulse is unipolar, the photo-EMF pulse consists of two parts with nearly equal amplitudes and opposite signs. The increase of sweep rate decreases the pulse duration, while the amplitude remains nearly constant. At large A chirped and damped oscillations appear.

The signals calculated for the regime with space-charge wave excitation have several features. The pulses generated at small chirp rates appear at the moments, when the interference pattern runs synchronously with the space-charge waves. The pulses of the photo-EMF excited at low chirp rates are unipolar in contrast to those in the diffusion regime.

3. Experimental results and discussion

The diffusion regime of signal excitation has been realized for this time [2]. Time dependencies of the TWM and photo-EMF signals were measured in photorefractive $Bi_{12}TiO_{20}$ and GaAs crystals (Fig. 2). The results of theoretical analysis and experiments generally match each other. The time of signal appearance, variation of the pulse duration and shape versus sweep rate are well described by the theory. However, some features such as pulse tails in GaAs and low pulse amplitude in $Bi_{12}TiO_{20}$ require more advanced theoretical approaches dealing with bipolar conductivity, high light absorption, vectorial coupling.

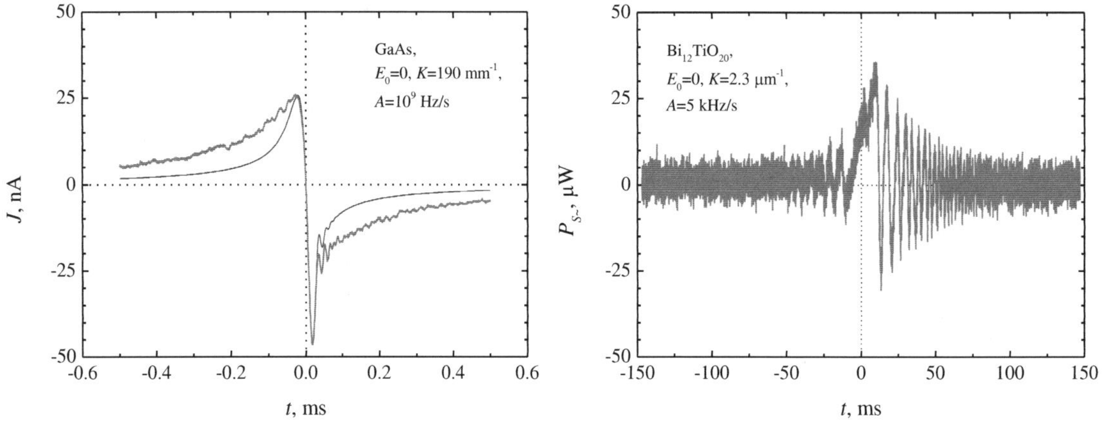

Figure 2. Oscillograms of the non-steady-state photo-EMF in GaAs and TWM signal in $Bi_{12}TiO_{20}$ excited by an accelerating interference pattern. λ=633 nm, I_0=240 mW/cm^2. Thin blue line shows approximation by Eq. (5) for τ_{sc}=12 μs.

The pulses appear at the moments when frequencies of the signal and reference beams coincide. This fact can be used in Doppler scanning velocimeter estimating the unknown velocity from the time delay of the detected pulses. Moreover, the duration and shape of the pulses provide information about chirp rate of the LFM signal, which can be caused by the uniformly accelerated motion of the object. This means that the velocity and acceleration can be measured simultaneously by detection of the only pulse, what is very important in investigations of fast processes. In this work we used frequency shifts and sweep rates in ranges $|\Delta f|$=0-10 MHz and $|A|$=0-1000 MHz/s. They would correspond to the following ranges of velocity and acceleration of the light reflecting object: $|V|$=0-3 m/s and $|a|$=0-300 m/s^2.

Acknowledgements
The authors acknowledge the financial support from the Russian Science Foundation (project 15-12-00027).

References
[1] Günter P and Huignard J P 1989 *Photorefractive materials and their applications II* (Berlin: Springer)
[2] Bryushinin M, Kulikov V, Sokolov I, Delaye P and Pauliat G 2014 *J. Opt. Soc. Am. B* **31** 723

IOP Conf. Series: Journal of Physics: Conf. Series **867** (2017) 012005 doi:10.1088/1742-6596/867/1/012005

Experimental and theoretical investigations of the gain dependence at two-wave interaction on the thickness and orientation angle of the $Bi_{12}GeO_{20}$ crystal

V V Shepelevich[1], **A V Makarevich**[1] **and S M Shandarov**[2]

[1] Department of Theoretical Physics and Applied Informatics, I.P. Shamyakin Mozyr State Pedagogical University, 28 Studencheskaya Str., Mozyr, 247760, Belarus
[2] Department of Electronic Devices, Tomsk State University of Control Systems and Radioelectronics, 40 Lenin Ave., Tomsk, 634050, Russia

E-mail: vasshep@inbox.ru

Abstract. Experimental studies of the dependence of object wave gain at two-wave interaction on the effective thickness of cubic photorefractive optically active crystal $Bi_{12}GeO_{20}$ were performed using only one $(\bar{1}\bar{1}0)$ -cut crystal sample. It is shown that the obtained experimental results can be satisfactorily theoretically interpreted taking into account the inverse piezoelectric and the photo-elastic effects in addition to the traditionally considered electro-optical one.

1. Introduction

Photorefractive sillenite crystals $Bi_{12}SiO_{20}$ (BSO), $Bi_{12}GeO_{20}$ (BGO) and $Bi_{12}TiO_{20}$ (BTO) have found application in many holographic devices. This is caused by the high photosensitivity of such crystals and other unique their properties. To increase the efficiency of practical use of these crystals, it is of interest to experimentally and theoretically study the gain dependence of an object light wave in two-wave interaction on the thickness of the crystalline sample.

2. Experimental method

We carried out this investigation for $(\bar{1}\bar{1}0)$-cut BGO crystal using only one sample 16 mm thick. The realization of such experiment has been possible with the use of trapezoidal intersection geometry of the reference and object light beams in the crystal. This geometry was first advanced in [1–3] also for studies of the gain in sillenite crystals. However, it is important to note that in these studies only such spatial orientations of the crystals with respect to the holographic grating vector **K** were considered for which this vector was perpendicular or parallel to the [001]-crystallographic direction.

This made it possible to eliminate the inverse piezoelectric and photoelastic effects (hereinafter it abbreviated as "piezoelectric effect") from the theoretical calculations for **K**||[001] and neglect the piezoelectric effect at **K**⊥[001]. Therefore, we considered other orientations of the BGO crystal relative to the plane of propagation of light beams and showed the importance of taking into account the piezoelectric effect. Note that the trapezoidal geometry was first used to experimentally study the dependence of *the diffraction efficiency* of holograms on the thickness of the BSO crystal in [4].

3. The experimental conditions and parameters of BGO crystal

During the experiments with BGO at a wavelength $\lambda = 632.8$ nm, the ratio of the intensities of object S and reference R light beams before entering the crystal I_S^0 / I_R^0 was approximately equal to 0.25. The Bragg angle φ_0 outside the crystal was 15°. The spatial orientation of the crystal relative to the plane of propagation of the light beams, determined by the angle θ, corresponded to [4], as well as the choice of the direction of the azimuth of the linear polarization Ψ_0 of the beams.

In the theoretical calculations, the following parameters of the BGO crystal were used: specific rotation of the plane of polarization $\rho = 363$ rad/m; refractive index $n = 2.54$; electro-optical coefficient $r_{41} = -3.4\times10^{-12}$ m/V; coefficients of elasticity $c_1 = 12.8\times10^{10}$ N/m^2, $c_2 = 3.05\times10^{10}$ N/m^2, $c_3 = 2.55\times10^{10}$ N/m^2; piezoelectric coefficient $e_{14} = 0.99$ C/m^2; photoelastic constants $p_1 = -0.136$, $p_2 = -0.103$, $p_3 = -0.091$, $p_4 = -0.0134$.

4. Obtained experimental results and their comparison with theoretical ones

The obtained experimental and theoretical results of studying the dependence of the object light wave gain γ on the effective thickness d of the BGO crystal, in particular, for θ = 70° and 250° at $\Psi_0 = 0$ and 90°, are shown in figure 1.

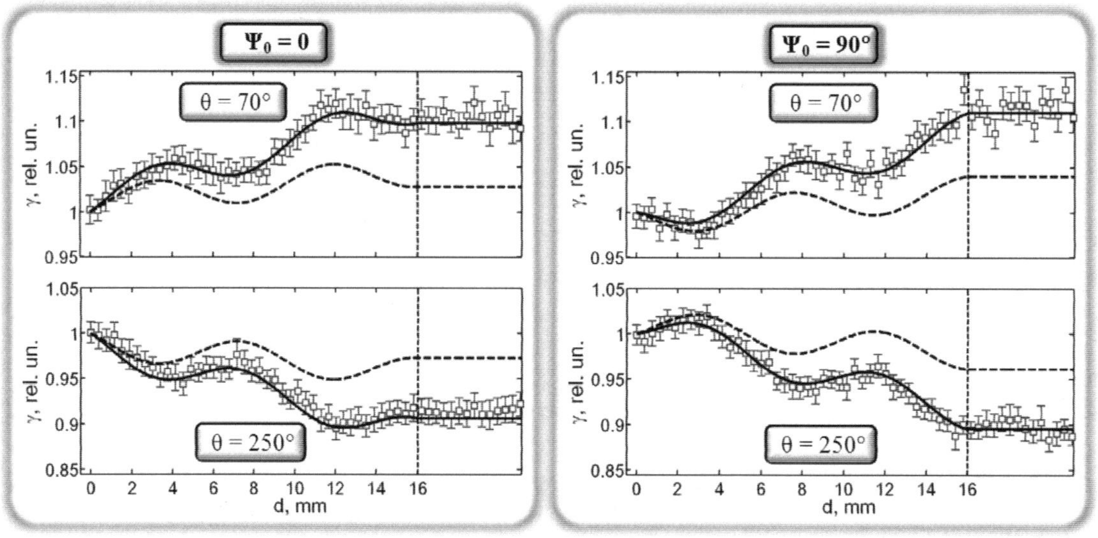

Figure 1. Dependences $\gamma(d)$ for the crystal of the BGO $(\bar{1}\bar{1}0)$-cut with $d_0 = 16$ mm: dashed line – theory for $e_{14} = 0$, solid line – theory for $e_{14} \neq 0$; □ – experiment

As can be seen from Fig. 1, only additional accounting of the piezoelectric effect leads to reach agreement between theoretical and experimental results.

5. Conclusion

Thus, the inclusion of piezoelectric effect leads to the most efficient use of the BGO crystal in holographic applications.

This effect should be taken into consideration in the theoretical determination of the optimal conditions of the two-wave interaction in the investigated crystal (for example, by choosing the optimal values of θ, Ψ_0 and d, at which the maximum gain of the object light wave is reached).

Acknowledgments
This work was supported by the Ministry of the Education of Belarus (project 1.2.01 of the State Program "Photonics and Opto- and Microelectronics), Ministry of the Education and Science of the Russian Federation and the Russian Foundation for Basic Research (project no. 16-29-14046-ofi_m).

References

[1] Shamonina E, Mann M, Ringhofer K H, Kiessling A and Kowarschik R 1996 *Opt. Quant. Electron.* **28** 25
[2] Shamonina E, Cedilnik G, Mann M, Kiessling A, Webb D J, Kowarschik R and Ringhofer K H 1997 *Appl. Phys. B.* **64** 49
[3] Shamonina E, Kamenov V P, Ringhofer K H, Cedilnik G, Kießling A, Kowarschik R and Webb D J 1998 *Opt. Comm.* **146** 62
[4] Makarevich A V, Shepelevich V V, Ropot P I, Naunyka V N and Shandarov S M 2015 *Techn. Phys. Lett.* **41** 942

Mid-Infrared photorefractivity with non-degenerate two-photon absorption

B Bourdon[1], F Freytag[1], M Imlau[1], A Shumelyuk[2] and S Odoulov[2]

[1] Department of Physics, Osnabrueck University, Barbarastrasse 7, D-49076 Osnabrueck, Germany
[2] Institute of Physics, NASU, Science Ave. 46 UA-03650 Kyiv, Ukraine

Corresponding author's e-mail address: bjbourdon@uni-osnabrueck.de

Abstract. Photorefractive grating recording with mid-infrared sub-picosecond light pulses (1.6 μm and longer wavelengths) might be possible in $LiNbO_3$ crystals if the interaction area is illuminated additionally with pulses of visible light. Two processes can enable the experimentally observed MIR recording: One is related to the direct photoexcitation of a free electron via nonlinear absorption of two photons with different frequencies, one in the MIR and one in the visible range. The alternative possibility consists in an enhancement, caused by the diffraction of a visible pulse, of a very weak space-charge grating seeded solely by MIR radiation.

1. Introduction

Until now the recording of photorefractive gratings with infrared laser light was reported with wavelengths not exceeding 1.55 μm [1-3]. We show in this contribution that - profiting from strongly non-degenerate two-photon or multiphoton absorption (see for example [4]) of a train of sub-picosecond pulses - it becomes possible to imprint the fringes of infrared (IR) light as a space-charge grating, when simultaneously illuminating the crystal homogeneously with a visible pulse. This new excitation scheme opens the era of MIR dynamic holography and may potentially be applied for MIR-pulse-characterization or even MIR-pulse-control in medical surgery.

2. Experimental Setup

A 1-mm thick y-cut crystal of $LiNbO_3$:Fe (0.1 wt.%) with optically finished input/output faces is illuminated by two IR beams with $\lambda = 1.6$ μm that impinge upon the sample (see figure 1a). These beams are generated by one of two tuneable OPA's each generating a pulse train with a 1 kHz repetition rate and a pulse duration of $\tau \approx 100$ fs, polarized perpendicular to the optical c-axis. The IR pulses with energies $E_{IR} \leq 25$ μJ are loosely focused ($r \approx 180$ μm) into the sample and superposed in time as well as in space. In addition, a train of visible pulses is synchronized with the IR pulses within the same crystal area; this beam is loosely focused, too, to cover completely the intersection area of the IR pulses. Finally, a cw beam of a He-Ne laser is sent to the intersection area of the sub-picosecond pulses at an angle that matches the Bragg diffraction condition from the inertial space charge grating that might be recorded by the IR beams in presence of the visible pulse. Both, the visible pulse train and the red beam are polarized normal to the $LiNbO_3$ c-axis in order to reduce the wavefront distortions because of optical damage. By adjusting the Bragg angle of incidence for the train of ultrashort visible pulses we get an extra tool for directly detecting the instantaneous two-photon absorption grating within every single pulse.

3. Results & Discussion

If two IR pulses and one visible pulse are present in the $LiNbO_3$ sample simultaneously, the diffracted red beam appears as it is shown in figure 1(a). The intensity of the diffracted beam increases with time and typically saturates within twenty seconds for an IR pulse energy of $E_{IR} \approx 10$ μJ. No diffraction is detected if the visible beam is absent or very weak. Figure 1(b) shows the saturated intensity of the diffracted red beam I versus the visible pulse energy E_{VIS} for repetition rate recording, indicating a pronounced nonlinear behavior for moderate energies: $I \sim E_{VIS}^4$; the dependence versus the IR pulse energy is also nonlinear. Thus we conclude that a photorefractive grating is recorded by common action of IR and visible light.

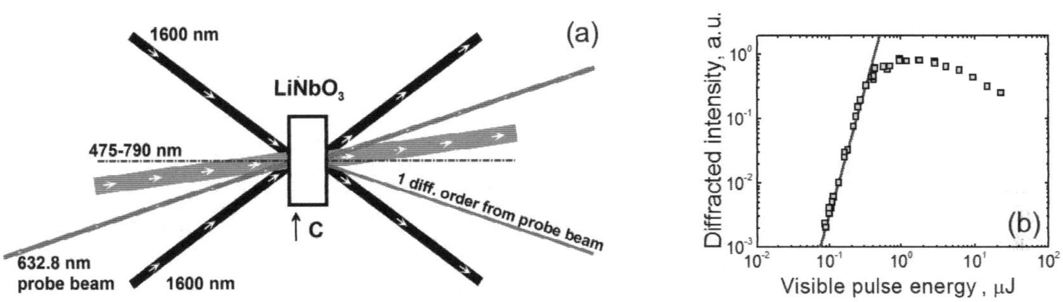

Figure 1, (a) Experimental geometry. Two infrared beams produce the fringes in $LiNbO_3$ sample ($2\theta_{IR} = 7.7°$) that are superposed to homogeneous illumination by a visible beam; altogether they produce an instantaneous two-photon absorption grating and trigger the formation of a long-living photorefractive grating. A Bragg-matched red beam ($\theta_{red} = 3°$) allows for monitoring the build-up of the photorefractive grating. (b) Saturated intensity of the diffracted red beam as a function of visible (510nm) pulse energy (total energy of two recording IR pulses 25 μJ, repetition rate 1 kHz).

The temporal delay of the green pulse with respect to the MIR pulses (advanced and retarded, whatever) prevents from formation of both, instantaneous and long-lasting gratings. This allows for excluding possible two-step excitation of free carriers from impurity or defect energy levels within the band gap. In other experiments it was established that the absorption of a MIR pulse increases in presence of a visible pulse and vice versa, the visible pulse absorption becomes larger in presence of MIR pulse. In such a manner the existence of strongly non-degenerate two photon absorption with MIR involved is supported.

Acknowledgments

We gratefully acknowledge financial support from the Deutsche Forschungsgemeinschaft (DFG) via projects IM37/5-2, INST 190/137-1 FUGG and INST 190/165-1 FUGG.

References

[1] Wesner M, Herden C, Kraetzig E, Kip D and Moretti P 2001 *Opt. Commun.* **188** 69
[2] Mosimann R, Marty P, Bach T, Juvalta F, Jazbinsek M, Günter P and Grabar A 2007 *Opt. Lett.* **32** 3230
[3] Beyer O, Breunig I, Kalkum F and Buse K 2006 *Appl. Phys. Lett.* **88** 051120
[4] Fishman D A, Cirloganu C M, Webster S, Padilha L A, Monroe M, Hagan D J and Van Stryland E W 2011 *Nat. Photon.* **5** 561

Extreme events induced by self-action of laser beams in dynamic nonlinear liquid crystal cells

S Bugaychuk[1,*], **A Iljin**[1] **and K Chunikhina**[2]

[1]Institute of Physics, National Academy of Sciences of Ukraine, 46 Prospect Nauki, Kiev 03039, Ukraine
[2]Faculty of Radiophysics, Electronics and Computer Systems, Taras Shevchenko National University, 4-G Prospect Glushkova, Kiev 03650, Ukraine

*e-mail: bugaich@iop.kiev.ua

Abstract. Optical extreme events represent a feature of nonlinear systems where there may emerge individual pulses possessing very high (or very low) intensity hardly probable statistically. Such property is being connected with the generation of solitons in the nonlinear systems. We carry out the first experiments for detection of extreme events during two-wave mixing with nonlinear dynamical liquid crystal (LC) cells. We investigate the statistics of the extreme events in dependence on relation between the duration of a laser pulse and the time characteristic of dynamic grating relaxation in LC cell. Our research shows that the self-diffraction of laser beams with a dynamical grating support the generation of envelope solitons in this system.

1. Introduction

Investigations of soliton dynamics and rogue waves formation are carrying out for many optical nonlinear systems including Kerr cavities with controllable feedback and photonic crystal fibers [1,2]. Wave self-diffraction in a dynamic medium with amplification and relaxation may be described the complex Ginzburg-Landau equation (CGLE) [3, 4] having solutions as dissipative solitons, breathers and rogue waves. The beam coupling in nonlinear liquid crystal (LC) cells represents a novel system for such studies. The present work is devoted to the search for optical extreme events and rogue waves during interaction of laser pulse series inside the LC medium. In specially chosen conditions depending on mutual relations between the periods of pulse series and the characteristic times of the LC system, we have detected the generation of unexpected behaviour for output both signal and pump beams. The extreme events excited in complex optical systems during dynamic holographic experiments presents interest for all-optical information processing, in particular, for security of optical information.

2. Experiment and discussion

A simple arrangement has been chosen for the LC cell. Two glass substrates covered by transparent ITO electrodes were set apart with ~30 μm gap and a dye-doped nematic LC E7 was injected in between. The dye Ethyl Red was used in the concentration of ~1 w.%. Application of the ac voltage (~100 V @ 1÷10 kHz) ensured the orientation of the LC director perpendicularly to the substrates and prevented its reorientation.

Conventional two-wave mixing technique was used for the dynamic grating induction in the cells. The laser beam from a Nd:YAG continuous laser (single-mode operation, λ=532 nm) was split into two beams, which converge on a sample to form interference pattern. The intensity of the pump beam was $I_p = 0.5\,\text{mW}$, of the signal beam $I_s = 0.58\,\text{mW}$. A chopper put into the signal beam provided square-shaped pulses of ~1/30 of the whole revolution period, which was controlled by a voltage value of a power source.

Recording of the dynamic grating was detected due to the appearance of diffracting orders, then the intensity the signal and pump beams at the output from the cell has been recoded. A fast and slow optical response has been observed. The fast process occurred at lower intensities (< 100 mW/cm^2) and had the characteristic times of grating recording (relaxation) about $5(100)\,\text{ms}$. It was attributed to the photo-induced modulation of the LC order parameter resulting from *trans-cis* isomerisation of dye molecules, which, in turn, caused the changes of refractive indices of the medium (Light-Induced Order Modification, LIOM-mechanism [5]). At higher pump intensities (up to 300 mW/cm^2) the collective reorientation of LC molecules took place in the LC bulk, with the characteristic times of the order of seconds.

We calculated the statistics of the intensities of the output pulse in dependence on the duration on the pulses in comparison with the time relaxation constant of the dynamical grating. In figure 1 we present the theoretical calculations with taking account of the actual experimental parameters. The theoretical model is based on self-diffraction of two waves on an existing local dynamic grating with arbitrary phase mismatch in the Bragg condition. The estimated value of the phase modulation of the dynamical grating in our cells is 0.2.

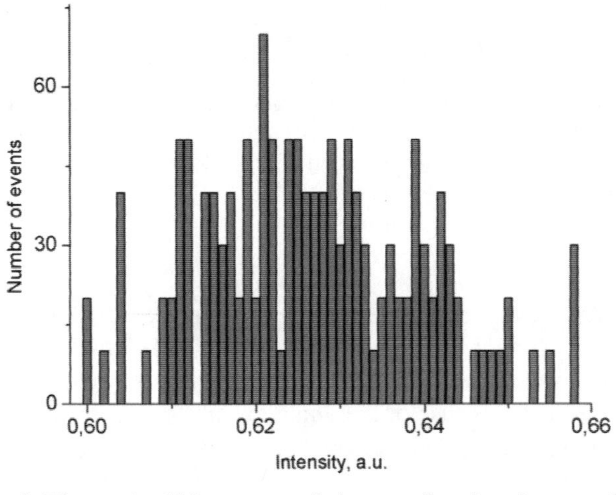

Figure 1. Theoretical histogram of changes for signal output intensity.

Our experimental and theoretical results show that the statistics of the appearance of high or low intensities at the output does not satisfy a classical probability theory followed from the linear systems. The careful classification of the extreme events in dependence on relation between the duration of the laser pulses and the time relaxation constant of the nonlinearity in the nonlinear LC cells are the subject of our report and further investigations.

References

[1] Sollin D R, Ropers C, Koonath P and Jalali B 2007 *Nature* **450** 1054
[2] Akhmediev N, Ankiewicz A and Taki M 2009 *Physics Letters A* **373** 675
[3] Bugaychuk S and Conte R 2009 *Phys. Rev. E* **80** 066603
[4] Bugaychuk S and Conte R 2012 *Phys. Rev. E* **86** 026603
[5] Iljin A 2011 *Mol. Cryst. Liq. Cryst.* **543** 143

IOP Conf. Series: Journal of Physics: Conf. Series **867** (2017) 012008 doi:10.1088/1742-6596/867/1/012008

Flexoelectric contribution to the phase demodulation by two-beam coupling on reflection and transmission gratings in cubic photorefractive crystals

S M Shandarov[1], N I Burimov[1], S S Shmakov[1], A O Zlobin[1], A A Shmidt[1], V V Shepelevich[2], A V Makarevich[2], Yu F Kargin[3], K Shcherbin[4]

[1]Tomsk State University of Control Systems and Radioelectronics, Tomsk 634050, Russia
[2]I.P. Shamyakin Mozyr State Pedagogical University, Mozyr 247760, Belarus
[3]Baikov Institute of Metallurgy and Material Sciences of the RAS, Moscow 119991, Russia
[4]Institute of Physics, National Academy of Sciences, 03680 Kiev, Ukraine

E-mail: stanislavshandarov@gmail.com

Abstract. We study the linear and quadratic signals of phase demodulation in holographic interferometers, which are realized at interaction of light waves on dynamic hologram of diffusion type in the cubic photorefractive crystals, with taking into account the flexoelectric contribution to the nonlinear response.

1. Introduction

The converse flexoelectric effect that consists of the appearance of elastic strain that are proportional to the gradient of the electric field provides the changes in refractive index of the crystal by photorefractive gratings due to photoelasticity, which are shifted by $\pi/2$ relative to ones connected with linear electro-optic effect [1]. Because of that this flexoelectric contribution having the magnitude varies inversely with fringe spacing Λ provides an observable linear component of phase-demodulation signal in holographic interferometers using contradirectional two-beam interaction on reflection grating of the diffusion type in the cubic photorefractive crystals [1 – 3]. In this report we consider the effect of flexoelectricity on the signal of phase demodulation in holographic interferometers on the base of codirectional and contradirectional variants of interaction on the holograms of diffusion type in the cubic photorefractive crystals.

2. Theory

The relative amplitudes of the first and second harmonics in the time-dependent intensity of output signal beam of holographic interferometer can be represented in many cases respectively as

$$M^{(1)}(\varphi_m) = 4\,J_0(\varphi_m)\,J_1(\varphi_m)\exp\left(\frac{\Gamma_E d}{2}\right)\sin\left(\frac{\Gamma_f d}{2}\right), \qquad (1)$$

$$M^{(2)}(\varphi_m) = 4\,\mathrm{J}_0(\varphi_m)\,\mathrm{J}_2(\varphi_m)\left[\exp\left(\frac{\Gamma_E d}{2}\right)\cos\left(\frac{\Gamma_f d}{2}\right) - 1\right], \quad (2)$$

where φ_m are the phase modulation amplitude of the signal beam, J_n is the n-th Bessel function, and Γ_E is the coupling coefficient descriptive of the conventional electro-optic contribution to the response. The coupling coefficient $\Gamma_f \sim f_{eff}$, where f_{eff} is effective flexoelectric coefficient, describing interaction on the grating component, which is due to flexoelectric and elasto-optic effects. For the coupling of left-circularly polarized pump beam with right-circularly polarized signal beam on the reflection grating in the (100)-cut crystals with symmetry group 23 the analytical expression for Γ_f was presented previously [1, 2]. For the pump and signal beams interacting on the reflection grating in the (111)-cut crystals both with symmetry group 23 at circular polarization of opposite signs and with symmetry group $\bar{4}3m$ at identical linear polarization we obtain

$$\Gamma_f^{(111)} = \frac{2\pi}{\lambda} n_0^3 \frac{3(p_{11} + 2p_{12} - 2p_{44}^E)}{c_{11} + 2c_{12} + 4c_{44}^E}\left(\frac{2\pi}{\Lambda}\right) E_{SC} f_{(111)}, \quad (3)$$

$$f_{(111)} = \frac{1}{3\sqrt{3}}\left(f_{1111} + f_{1122} + f_{2211} + 2f_{2323} + 2f_{2332}\right), \quad (4)$$

where λ is the light wavelength, n_0 is the refractive index for an unperturbed crystal, E_{SC} is the effective space-charge field of hologram [1, 2], c_{mn}, p_{ij} and f_{ijkl} are the elastic modulus and elasto-optic and flexo-coupling coefficients, respectively.

3. Experiment

The results of experimental investigation of flexoelectric contribution to the phase demodulation at contradirectional interaction in the cubic crystals $Bi_{12}TiO_{20}$ and CdTe were presented before [1 – 3]. The phase demodulation at the codirectional interaction in the (110)-cut $Bi_{12}SiO_{20}$ crystals for light waves with identical polarizations was studied for holograms with the spatial period from 0.66 to 1.03 μm. From the experimental data for photorefractive gratings with a grating vector directed along the [001] crystallographic axis we estimated the values of flexo-coupling coefficient of $Bi_{12}SiO_{20}$ crystals as $f_{1111} \sim 10^{-8}$ C/m.

4. Conclusion

We have considered the phase demodulation in holographic interferometers using the holograms of diffusion type in the cubic photorefractive crystals with taking into consideration the flexoelectric contribution to the response.

Acknowledgments

This work was supported by the Ministry of the Education of Belarus (project 1.2.01 of the State Program "Photonics and Opto- and Microelectronics), the State Assignment of the Ministry of Education and Science of Russian Federation for 2017 and the RFBR (grant No. 16-29-14046-ofi_m).

References

[1] Shandarov S M, Smakov S S, Burimov N I, Syuvaeva O S, Kargin Yu F, Petrov V M 2012 *JETP Letters* **95** 618

[2] Shandarov S M, Smakov S S, Zuev P V, Burimov N I, Kargin Yu F, Shepelevich V V, Ropot P I, Gudelev V G 2013 *J. Opt. Technol.* **80** 409

[3] Shandarov S M, Burimov N I, Zlobin A O, Bykov V I, Mel'nik K P, Shepelevich V V, Makarevich A V, Kargin Yu F, Shcherbin K 2015 *Proc. PR15 – Photorefractive Photonics (Villars, Switzerland)*

IOP Conf. Series: Journal of Physics: Conf. Series **867** (2017) 012009 doi:10.1088/1742-6596/867/1/012009

Optical mapping of oscillatory stresses in transparent media

Mikhail Bryushinin, Vladimir Kulikov, Alexander Petrov, Igor Sokolov

Ioffe Institute, Politekhnicheskaya 26, 194021 St. Petersburg, Russia

i.a.sokolov@mail.ioffe.ru

Abstract. The phase-modulated optical signal produced by the light propagation through the stressed transparent media is detected using the non-steady-state photoelectromotive force technique. The mechanical system including the glass plate and piezoelectric transducer demonstrates the resonant behavior in the vicinity of 100 kHz. The measured distribution of the optical phase is a bell-shaped surface for these frequencies. The simulation of the piezooptic response shows the quali-tative coincidence with the experimental data and provides the maps of the stress field. The characterization of the β-Ga_2O_3 adaptive detector is performed for the light wavelength $\lambda = 532$ nm.

1. Introduction

The detection of phase-modulated optical signals using adaptive detectors is usually carried out using the vibrometric scheme, where the light phase incursion is created by the reflection from an oscillating object [1]. Such an additional phase can obviously appear in the light beam propagating in a stressed transparent material as well. In this work we apply the non-steady-state photo-EMF technique to the detection of phase-modulated optical signals produced by the oscillating stresses in the transparent material.

2. Experimental procedure and results

The experiments with the detection of oscillatory stresses are carried out with the arrangement shown in Fig. 1. One of the beams passes through the transparent glass plate PM-15. The plate is made of glass K8 (BK7 analog) and has the diameter of 30.3 mm and thickness of 15.87 mm. The front and back surfaces have the optical quality. The lateral (cylindric) surface has two diametrically arranged 6.6 mm flats, which were used for clamping. The oscillatory stresses in this plate are created by the piezoelectric transducer. The adaptive detector is made of monoclinic β-Ga_2O_3 crystal and has dimensions 2.00×2.15×1.35 mm along crystallographic directions [100], [010] and direction perpendicular to plane (001).

First, we have calibrated our adaptive detector by measuring the dependencies of the signal amplitude versus temporal and spatial frequencies. The frequency transfer function looks like that of differentiating RC-circuit with the cut-off frequency of 13 Hz (for the light intensity I_0=0.4 W/cm^2). The spatial frequency K=4.0 μm^{-1} chosen for the further experiments with the glass plate is very close to its optimum value maximizing the signal.

The photo-EMF signal produced by the stress oscillations in the glass plate has the frequency dependence with resonant peaks at 93 and 119 kHz. The resonant behavior is associated with the mechanical resonances of the system including the glass plate, piezoelectric transducer and clamp.

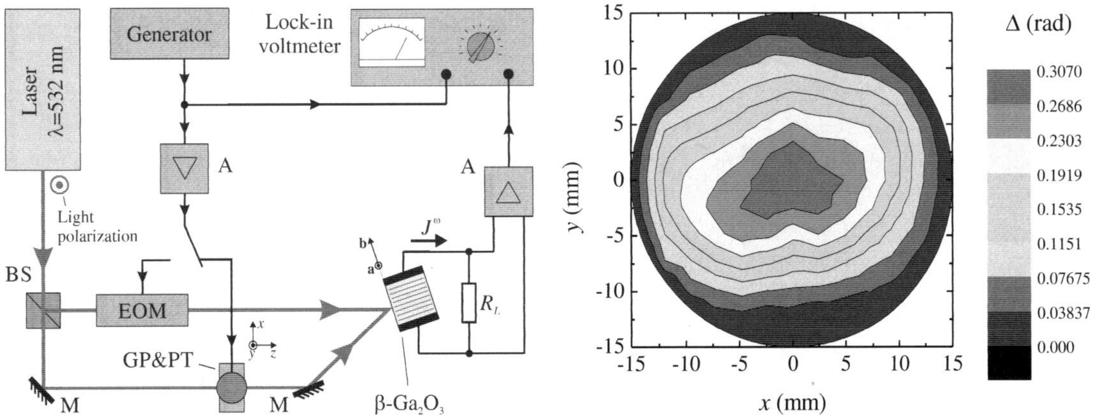

Figure 1. Experimental setup (on the left) and distribution of the phase modulation amplitude across the glass plate surface (on the right). EOM is the electrooptic modulator, GP&PT are the glass plate and piezoelectric transducer.

The glass plate has the rather complicated geometry, and one can expect that the mechanical stresses, deformations and resulting amplitude of phase modulation non-uniformly distribute across the plate surface. We have measured this distribution (Fig. 1) by moving the clamp with the glass plate and transducer in the directions perpendicular to the light beam. The measurements are performed at 93.5 kHz, i.e. at the first resonant frequency. As seen, the distribution has the nearly radial symmetry with the maximum at the center of the plate.

3. Simulation

The second part of the work contains the simulation of the light phase distribution in the stressed glass plate. The analytic approach for such a problem requires solution of motion equation for elastic medium. Since the cylinder has the two flats, and its height is finite, the analytic approach seems too difficult. Besides, the obtained solution would be valid only for this geometry, which was chosen rather arbitrarily. So, the numerical solution is much more reasonable, and we use the structural mechanics module of COMSOL Multiphysics software to cope with this problem. The geometry of the problem is the same as that used in the experiment. The bottom flat of the cylinder is fixed, the top flat oscillates with amplitude 1.5 nm and frequency 104 kHz. Figure 2 presents the diagonal components of the stress tensor σ calculated for the cross section at $z=d/2$, i.e. at half height of the cylinder. The analogous distributions of these values at different z including cylinder faces were calculated. They slightly differ from those presented in Fig. 2, so in the further analysis we can consider stresses σ_{11}, σ_{22} and σ_{33} as functions of coordinates x and y but not z.

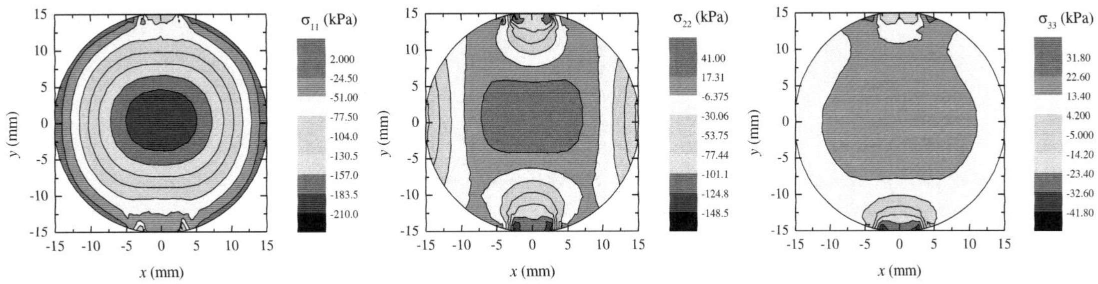

Figure 2. Calculated spatial distributions of the diagonal components σ_{11}, σ_{22}, σ_{33} of the stress tensor.

Then we proceed with the calculation of the light phase increment arising via piezooptic effect. These calculations are nearly the same as those for the electrooptic effect, except the tensor of dielectric impermeability is defined in a different way: $\xi_{ij}=\pi_{ijkl}\sigma_{kl}$, where π_{ijkl} is the tensor of piezooptic coefficients. If the stresses and corresponding phase increments are rather small, then the desired amplitude of phase modulation in the light wave polarized along *y*-axis is expressed as

$$\Delta(x,y) = -\frac{2\pi}{\lambda} n_0 d \left[\left(\frac{n_0^2 \pi_{12}}{2} + \frac{\nu}{E} \right) \sigma_1(x,y) + \left(\frac{n_0^2 \pi_{11}}{2} + \frac{\nu}{E} \right) \sigma_2(x,y) + \left(\frac{n_0^2 \pi_{12}}{2} - \frac{1}{E} \right) \sigma_3(x,y) \right], \quad (1)$$

where n_0 is refractive index, E is the Young's modulus, ν is the Poisson's ratio, and reduced notation of tensors is used. After substitution of the calculated stress tensor distributions (Fig. 2) we obtain the map of the light phase amplitude [2]. The calculated distribution of the phase modulation amplitude is a bell-shaped surface, its maximum is located in the vicinity of the plate's center, as it was observed in the experiment (Fig. 1).

Let us estimate the sensitivity of the developed technique. The minimal detectable amplitude of phase modulation is of $\Delta_{min}=1.6\times10^{-3}$ rad for the power of the signal beam P_s=1 mW, detection bandwidth δf=1 Hz and load resistor R_L=300 kΩ. Since the piezooptic coefficients π_{ij} are of the same order of magnitude and the Poisson's ratio ν is of order of unity, we can roughly estimate the minimal detectable amplitude of the stress oscillation:

$$\sigma_{min} \sim \left[\frac{2\pi}{\lambda} n_0 d \left(\frac{n_0^2 \pi_{ij}}{2} + \frac{\nu}{E} \right) \right]^{-1} \Delta_{min} = (1-2) \times 10^3 \text{ Pa}. \quad (2)$$

The conventional holographic methods employ the counting of the interference fringes in the pattern. The phase step between adjacent bright and dark fringes is of π, and this value is much higher than minimal detectable amplitude of our sensor. The better sensitivity of the adaptive detector mainly results from the narrow detection bandwidth, which is of order of 1 Hz in our setup and can easily be decreased to 0.01 Hz and lower using modern lock-in voltmeters.

Acknowledgments
The authors acknowledge the financial support from the Russian Science Foundation (project 15-12-00027).

References

[1] Stepanov S I, Sokolov I A, Trofimov G S, Vlad V I, Popa D and Apostol I 1990 *Opt. Lett.* **15** 1239
[2] Bryushinin M A, Sokolov I A and Balbashov A M 2017 *Appl. Opt.* **56** 170

IOP Conf. Series: Journal of Physics: Conf. Series **867** (2017) 012010 doi:10.1088/1742-6596/867/1/012010

Continuous-wave optical parametric oscillation tunable up to 8 µm wavelength

Ingo Breunig[1,2], Josef Urban Fürst[1], Kevin Hanka[1], and Karsten Buse[1,2,*]

[1] Laboratory for Optical Systems, IMTEK, University of Freiburg, Georges-Köhler-Allee 102, 79110 Freiburg, Germany

[2] Fraunhofer Institute for Physical Measurement Techniques IPM, Heidenhofstraße 8, 79110 Freiburg, Germany

E-mail: *`karsten.buse@ipm.fraunhofer.de`

Abstract. We demonstrate the first cw OPO emitting mid-infrared light at wavelengths up to 8 µm. This device is based on a 3.5-mm-diameter whispering gallery resonator made of silver gallium selenide ($AgGaSe_2$) pumped by a compact distributed feedback laser diode emitting light at 1.57 µm wavelength. Phase-matching is achieved for a c-cut resonator disk pumped with extraordinarily polarized light at this wavelength. The oscillation thresholds are in the mW region, while the output power ranges from 10 to 800 µW. Wavelength tuning is achieved via changing the radial mode number of the pump wave and by changing the resonator temperature. Simulations predict that whispering gallery OPOs based on $AgGaSe_2$ with diameters around 2 mm can generate idler waves exceeding 10 µm wavelength

1. Introduction

Continuous-wave optical parametric oscillators (cw OPOs) convert cw pump light at the wavelength λ_p to signal and idler light at $\lambda_\mathrm{s,i}$in a non-centrosymmetric crystal such that photon-energy conservation is fulfilled, i.e. $1/\lambda_\mathrm{p} = 1/\lambda_\mathrm{s} + 1/\lambda_\mathrm{i}$. Fundamentally, cw OPOs can be tuned to every output wavelength as long as it is larger than the one of the pump wave. This great potential of wavelength tunability is combined with narrow-linewidth emission and even with watt-level output powers employing conventional mirror-based systems. Thus, they are ideally suited for high-resolution laser spectroscopy [1]. A practical limit for the tuning range of cw OPOs is given by the transparency of the nonlinear-optical crystal they are based on. Operating them at wavelengths beyond the transparency range one will face significantly higher oscillation thresholds and lower conversion efficiencies.

The great majority of cw OPOs is based on oxide crystals. In these materials, the upper wavelength limit of the transparency range is given by multi-phonon resonances starting around 4.5 µm. Consequently, the mid-infrared tuning range of cw OPOs has been limited to wavelengths below 5.5 µm for more than half of a century [2].

Here, we present a continuous-wave optical parametric oscillator based on a whispering gallery resonator made of silver gallium selenide ($AgGaSe_2$). This material can be considered as transparent between 0.8 and 17 µm wavelength [3]. Thus, it is suitable for the generation of highly tunable mid-infrared light. For whispering gallery OPOs, the above mentioned photon-energy conservation has to be supplemented by the phase-matching condition $m_\mathrm{p} =$

$m_s + m_i + \Delta m$, relating the azimuthal mode numbers $m_{p,s,i}$ (number of field maxima along the propagation distance) of the interacting waves with the phase-matching number Δm. For c-cut $AgGaSe_2$, as in our approach, we find $\Delta m = \pm 2$ [4].

2. Experimental setup and results

The setup sketched in Fig. 1a,b and described in detail in Ref. [5] by Meisenheimer et al. comprises a disk resonator with 3.5 mm diameter and a coupling prism made of silicon. The pump light at 1.57 µm wavelength provided by a compact diode laser is converted in the resonator to signal and idler light. The latter two and the remaining pump light are coupled out via the prism and analyzed with respect to powers and wavelengths.

At pump powers exceeding the oscillation threshold of typically some mW, we observe optical parametric oscillation. By addressing pump resonances belonging to the different mode numbers q_p (number of field maxima along the radial direction), different branches of signal and idler wavelengths between 2 and 8 µm can be selected. Temperature variation enables 100-nm-wide wavelength tuning within one branch of the idler light (see Fig. 1c). The measured tuning branches are in good accordance with theoretical predictions. For a resonator diameter of 2 mm, we expect to extend the tuning range even beyond 10 µm wavelength. At 10 mW pump power, the output power P_s of the signal wave ranges from 100 to 800 µW, depending on the selected branch, i. e. the selected parametric process. For the idler wave, we estimate the output power by using the relation $P_i = (\lambda_s/\lambda_i)P_s$. This is valid if both generated waves are coupled out of the resonator equally. The resulting values are in the range between several 10 and several 100 µW. They can be considered as lower limits for the idler power. Due to the longer wavelength, the coupling efficiency of the idler wave will be stronger than the one of the signal wave.

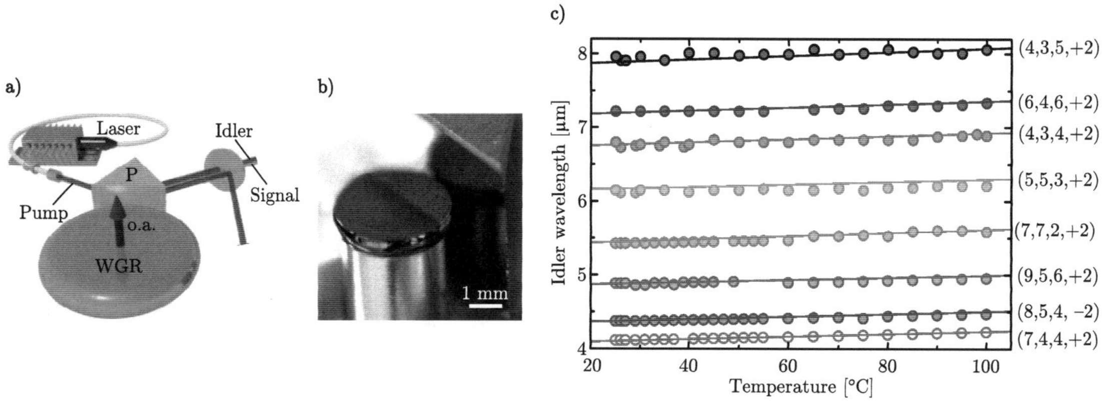

Figure 1. Whispering gallery optical parametric oscillator generating mid-infrared light. a) Sketch of the experimental setup comprising the resonator (WGR) and the coupling prism (P). The arrow indicates the optic axis (o.a.) of the crystal. b) Photograph of the resonator placed close to the coupling prism. c) Wavelength tuning of the idler light. The simulated tuning branches (solid lines) are labeled with the respective radial mode numbers $q_{p,s,i}$ and the phase-matching number Δm as $(q_p, q_s, q_i, \Delta m)$.

3. Conclusion

We consider these results to be a first step towards establishing continuous-wave optical parametric oscillators as light sources in the mid-infrared range beyond 5.5 µm wavelength. Further improvements with regard to output power and wavelength tuning are expected to be achieved with recently improved materials, e.g. orientation-patterned gallium arsenide and gallium phosphide.

IOP Conf. Series: Journal of Physics: Conf. Series **867** (2017) 012010 doi:10.1088/1742-6596/867/1/012010

References

[1] Vainio M and Halonen L 2015 *Phys. Chem. Chem. Phys.* **18** 4266
[2] Breunig I, Haertle D and Buse K 2011 *Appl. Phys. B* **105** 99
[3] Nikogosyan D N 2005 *Nonlinear optical crystals: A complete survey* (New York: Springer)
[4] Breunig I 2016 *Laser & Photon. Rev.* **10** 569
[5] Meisenheimer S K, Fürst J U, Buse K and Breunig I 2017 *Optica* **4** 189

IOP Conf. Series: Journal of Physics: Conf. Series **867** (2017) 012011 doi:10.1088/1742-6596/867/1/012011

High-order optical nonlinearities in nanocomposite films dispersed with semiconductor quantum dots at high concentrations

Yasuo Tomita[1], Shun-suke Matsushima[1], Ryu-ichi Yamagami[1], Taka-aki Jinzenji[1], Shohei Sakuma[1], Xiangming Liu[1,2], Takuya Izuishi[1], and Qing Shen[1]

[1] Department of Engineering Science, University of Electro-Communications, Tokyo 182-8585, Japan

[2] Research Center of Laser Fusion, Chinese Academy of Engineering Physics, Mianyang 621900, China

E-mail: ytomita@uec.ac.jp

Abstract. We describe the nonlinear optical properties of inorganic-organic nanocomposite films in which semiconductor CdSe quantum dots as high as 6.8 vol.% are dispersed. Open/closed Z-scan measurements, degenerate multi-wave mixing and femtosecond pump-probe/transient grating measurements are conducted. It is shown that the observed fifth-order optical nonlinearity has the cascaded third-order contribution that becomes prominent at high concentrations of CdSe QDs. It is also shown that there are picosecond-scale intensity-dependent and nanosecond-scale intensity-independent decay components in absorptive and refractive nonlinearities. The former is caused by the Auger process, while the latter comes from the electron-hole recombination process.

1. Introduction

Semiconductor quantum dots (QDs) that confine optically excited electron-hole pairs in all three space dimensions are of considerable interest in semiconductor physics and nonlinear optics [1, 2, 3]. In 2005 we suggested the possibility of optical bistable/multi-stable operations with nonlinear photonic lattice structure holographically recorded in a semiconductor QD dispersed nanocomposite film [4]. Later, we successfully demonstrated the holographic fabrication of fixed Bragg grating structure in a photopolymerizable nanocomposite film dispersed with CdSe QDs at the concentration up to 0.34 vol.% [5]. Such holographic pattering was made possible by holographic assembly of nanoparticles [6], resulting in the diffraction efficiency near 100%. We also reported the nonlinear optical response of uniformly cured nanocomposite films dispersed with CdSe QDs at concentrations as high as 0.91 vol.% [7]. Such a photopolymerizable nanocomposite material, the so-called photopolymerizable nanoparticle-polymer composite (NPC) [8], is prepared very differently from conventional inorganic nanocomposite materials that are generally fabricated by deposition/sputtering methods [9]. These nanocomposites can take advantage of the local field effect and the nanostructuring properties to control the linear optical, laser gain and nonlinear optical properties [9, 10]. Because of the syrup form of photopolymerizable NPCs they can be coated on any surfaces of substrate materials and

IOP Conf. Series: Journal of Physics: Conf. Series **867** (2017) 012011 doi:10.1088/1742-6596/867/1/012011

can also be infiltrated in nanophotonic device structures. In this summary we describe the fabrication of photopolymerizable inorganic-organic nanocomposite films, as a nonlinear optical version of NPCs, at CdSe QD concentration as high as $\sim$ 6.8vol.%, approximately 7 times higher than our previous report [7]. We also describe the transient and steady-state nonlinear optical properties of the uniformly cured nanocomposite films that exhibit the cascaded fifth-order optical nonlinearity. We perform Z-scan measurements, degenerate multi-wave mixing (DMWM) and femtosecond pump-probe measurements to evaluate absorptive and refractive nonlinearities in the transient and steady-state responses.

2. Experiments

2.1. Material preparation

We employed the aqueous synthetic method [5] to prepare colloidal CdSe QDs in water [see Fig. 1(a)]. The average core size of the synthesized CdSe QDs coated with 2-(dimethylamino)ethanethiol was estimated to be approximately 3 nm by a transmission electron microscope. It was followed by the extraction of CdSe QDs into an ionic liquid monomer possessing one methacryloyl group. The addition of colloidal CdSe QDs to the ionic liquid monomer resulted in the liquid-liquid phase separation by which the aqueous solution appeared on the monomer layer as shown in the upper photograph of Fig. 1(b). The CdSe QDs were extracted from the aqueous solution to the ionic liquid monomer layer by stirring action [see the lower photograph of Fig. 1(b)]. Such a repeated extraction procedure finally gave the CdSe QD- ionic liquid monomer mixture at the concentration of CdSe QDs as high as $\sim$10 vol.% [see Fig. 1(c)]. Visible photoinitiator (Irgacure 784, Ciba) was then added to the mixture for green-light sensitization. In order to obtain high crosslinking networks during polymerization, we also added an aliphatic urethane hexaacrylate monomer (Ebecryl 8301, Cytec) as a co-monomer to the mixture. Such mixed syrup was cast on a glass substrate with a polyester-film spacer and was covered with another glass substrate. It was then uniformly cured by green LED at 532 nm for our nonlinear optical measurements [see Fig. 1(d)]. The cured nanocomposite film exhibited the quantum-confined excitonic transition peak at 489 nm, independently of concentrations of CdSe QDs as shown in Fig. 2. Note that it is also possible to record a thick volume (Bragg) grating near 100% diffraction efficiency in such a uncured nanocomposite film [5, 8].

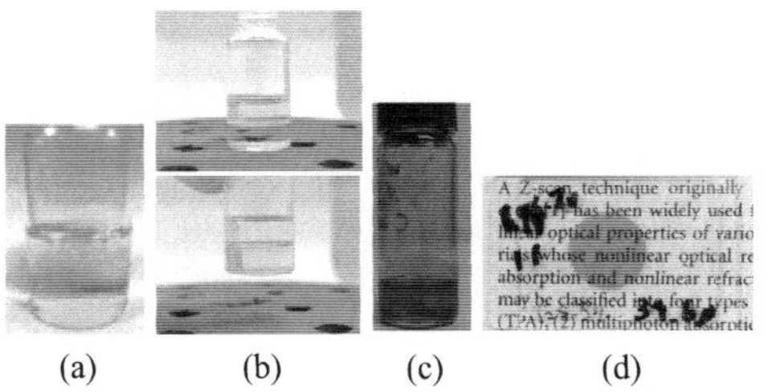

Figure 1. (a) Aqueous colloidal CdSe QDs; (b) Extraction of CdSe QDs from the aqueous CdSe QD solution (the yellowish layer) to an ionic liquid monomer (the transparent layer); (c) Photopolymerizable monomer syrup after repeated extraction of CdSe QDs; (d) A cured nanocomposite film sample dispersed with CdSe QDs.

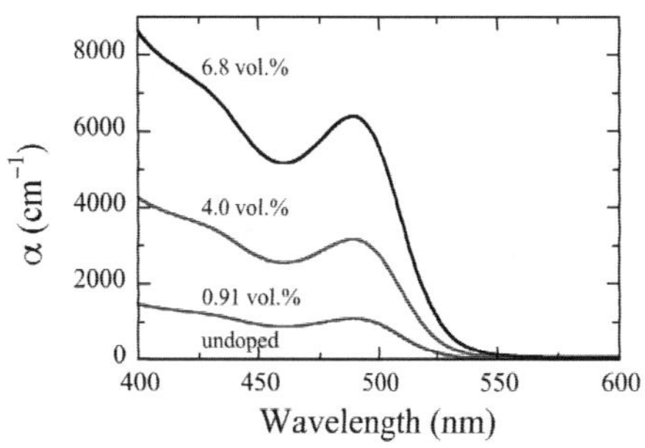

Figure 2. Spectral linear absorption coefficients α of cured nanocomposite film samples at various CdSe QD concentrations.

2.2. Nonlinear optical measurements

A frequency-doubled, passively and actively mode-locked picosecond Nd:YAG laser operating at a wavelength of 532 nm and at a repetition rate of 10 Hz was used in Z-scan and DMWM experiments. The normalized transmittance $T_{\mathrm{norm}}(z)$, defined as an energy ratio of a detected pulse to an incident pulse normalized by the linear transmittance, was evaluated as a function of the film sample's position z measured from the incident beam focus. Individual incident pulses at a given pulse intensity I_0 (measured at z=0) were selected within $\pm3\%$ errors, and averaging over tens of detected pulses was done to avoid unwanted errors in $T_{\mathrm{norm}}(z)$. Figure 3 shows $T_{\mathrm{norm}}(z)$ as a function of z for a cured film sample dispersed with 6.8 vol.% CdSe QDs at I_0=100 MW/cm^2. A generic theoretical model [11, 12] taking both saturable absorption for a homogeneous broadening two-level system [13] and third- and fifth-order nonlinear refraction into account was used to perform curve fitting to the measured data. We found that the effective third- and fifth-order nonlinear refraction coefficients $(n_2)_{\mathrm{eff}}$ and $(n_4)_{\mathrm{eff}}$ were found to be -1.5×10^{-2} cm^2/GW and $+7.0\times10^{-2}$ cm^4/GW^2, respectively, with the saturation intensity of 0.17 GW/cm^2. The origin of the observed nonlinear transparency is considered to be transient bleaching of the excitonic transitions as a result of the state filling effect [1] in CdSe QDs as observed by pump-probe spectroscopic measurements [14, 15]. We also found that a measured dependence of $(n_4)_{\mathrm{eff}}$ on the fill fraction f_i of CdSe QDs was well fitted to the following form of a quadratic dependence on f_i:

$$(n_4)_{\mathrm{eff}} = f_i|L|^4L^2(n_4)_{\mathrm{CdSe}} + f_i^2|L|^4L^2(n_2)^2_{\mathrm{CdSe}} - f_i(1-f_i)|L|^4L\left[2L^2(n_2)^2_{\mathrm{CdSe}} + |L|^2|(n_2)_{\mathrm{CdSe}}|^2\right],$$

where $(n_2)_{\mathrm{CdSe}}$ and $(n_4)_{\mathrm{CdSe}}$ are the third- and the fifth-order nonlinear refraction constants of a CdSe QD, respectively, and L is the local-field correction factor of Maxwell-Garnett composites [9, 10]. The first term in the right-hand side of the above equation corresponds to the direct fifth-order contribution, while the second and third terms correspond to the macroscopic and microscopic cascaded ones from the third-order nonlinearity, respectively [16, 17].

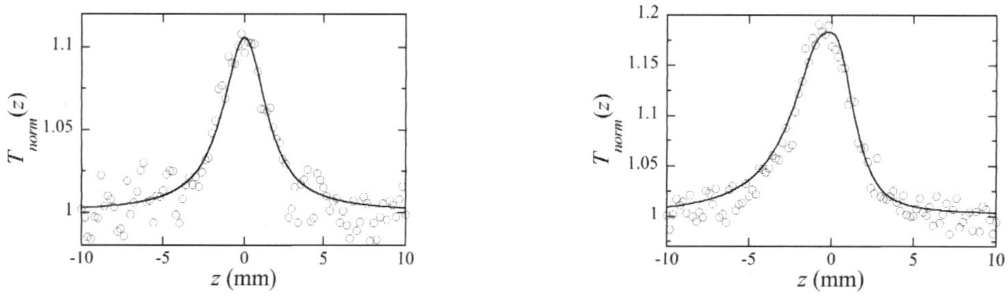

(a) (b)

Figure 3. (a) Open- and (b) closed-aperture Z-scan data for uniformly cured nanocomposite film sample. The solid curves are the least-squares fits of the theoretical models [11, 12] to the data.

Transmitted and diffracted beam patterns in DMWM measurements are shown in Figs. 4(a) and 4(b) for incident pump and probe beams at parallel and orthogonal polarizations, respectively. It can be seen from Fig. 4(a) that nonlinear diffraction originated from the third- and fifth-order optical nonlinearities take place for co-polarized pump and probe beams. The existence of the fifth-order nonlinearity is consistent with our Z-scan measurement described above. The observed third-order nonlinearity with cross-polarized pump and probe beams [see Fig. 4(b)] results from the non-zero component of the third-order nonlinear optical susceptibility tensor $\chi^{(3)}_{1221}$ of centro-symmetric media such as our nanocomposite films.

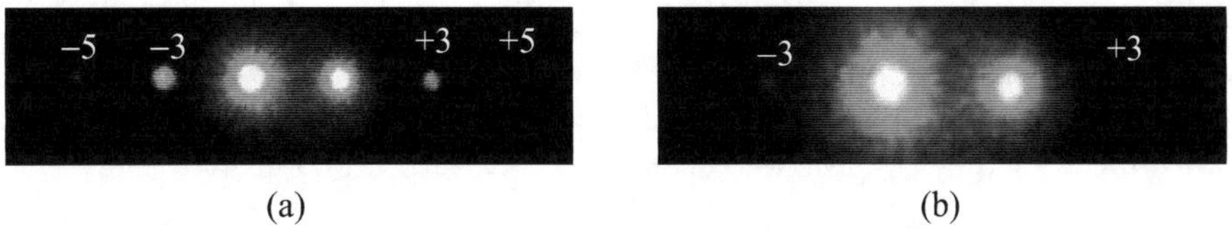

Figure 4. Transmitted and diffracted beams from a uniformly cured nanocomposite film sample in the DMWM setup with incident pump and probe beams at (a) parallel and (b) orthogonal polarizations.

We also performed femtosecond pump-probe measurements of absorptive and refractive nonlinearities induced in uniformly cured nanocomposite film samples. Figures 5(a) and 5(b) show the decay dynamics of the induced nonlinear refraction and transparency, respectively, in a nanocomposite film sample dispersed with 6.8 vol.% CdSe QDs, by means of the femtosecond transient grating method [18]. It can be seen that there are picosecond-scale intensity-dependent and nanosecond-scale intensity-independent decay components in the nonlinear refraction and transparency data. The former is caused by the Auger process as observed in colloidal CdSe QDs [19], while the latter comes from the electron-hole recombination process. Figure 5(c)

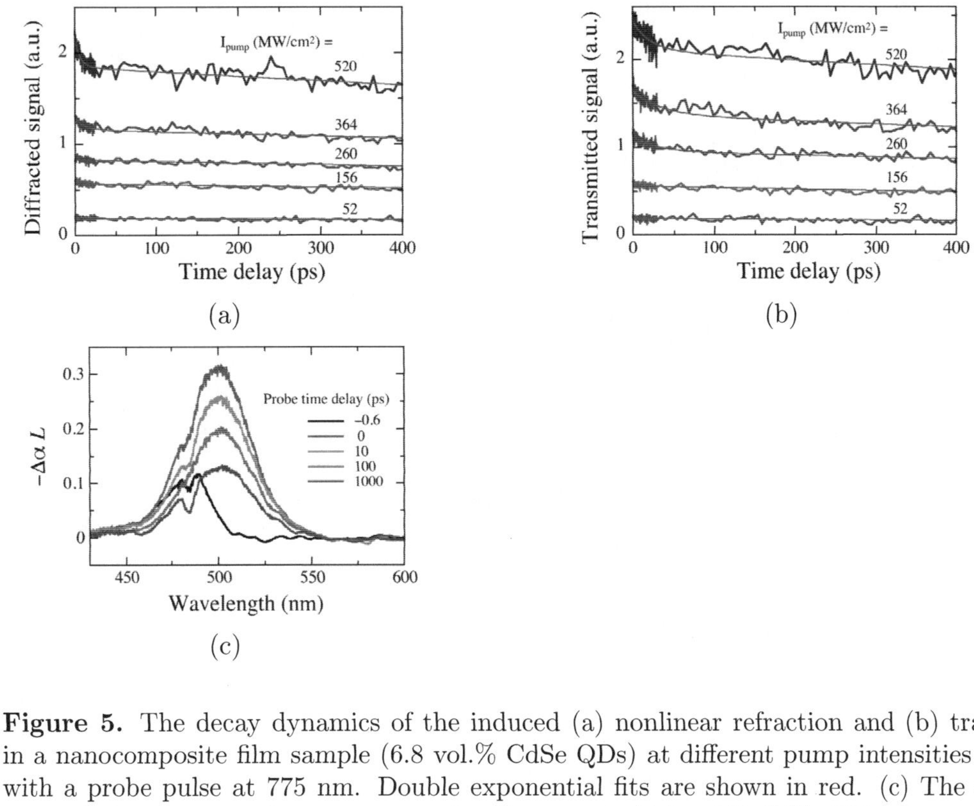

Figure 5. The decay dynamics of the induced (a) nonlinear refraction and (b) transparency in a nanocomposite film sample (6.8 vol.% CdSe QDs) at different pump intensities at 488 nm with a probe pulse at 775 nm. Double exponential fits are shown in red. (c) The differential transmission spectra of a nanocomposite film sample (0.9 vol.% CdSe QDs) for various probe time delays, where $\Delta\alpha$ is a change in α and L is the film thickness.

shows differential transmission spectra as a function of probe time delay for a nanocomposite film sample dispersed with 0.9 vol.% CdSe QDs at a 488-nm pump intensity of 260 MW/cm^2. It can be seen that saturable absorption takes place in the broadband spectral region around the first quantum-confined electron-hole transition wavelength as similar to results seen in Figs. 3(a) and 5(b).

3. Conclusion

We have studied the nonlinear optical properties of uniformly cured inorganic-organic nanocomposite films dispersed with CdSe QDs at high concentrations. We found that CdSe QD dispersed nanocomposite films exhibits negative third- and positive fifth-order nonlinear refraction with saturable absorption. We have also confirmed the fifth-order optical nonlinearity by DMWM. The observed quadratic dependence of the fifth-order nonlinearity on f_i indicates a contribution from the macroscopic and microscopic cascaded processes of the third-order effect in composite films. Such observed high-order nonlinear responses are interesting from a fundamental viewpoint of optical nonlinearities in nanocomposite structures [9]. In addition, aside from their straightforward applications of ultrafast switching, limiting and processing by film and nanophotonic device structures, the observed fifth-order nonlinearity may find applications in quantum-information science and manipulations of nonlinear optical pulses such as light bullets and spinning/multi-dimensional optical solitons.

References

[1] Bányai L and Koch S W 1993 *Semiconductor Quantum Dots* (Singapore: World Scientific)
[2] Haug H and Koch S W 2009 *Quantum Theory of the Optical and Electronic Properties of Semiconductors 5th ed* (Singapore: World Scientific)
[3] Banfi G P, Degiorgio V and Ricard D 1998 *Adv. Phys.* **47** 447
[4] Tomita Y 2005 *OSA Trends in Optics and Photonics Series (TOPS)* **99** 274
[5] Liu X, Tomita Y, Oshima J, Chikama K, Matsubara K, Nakashima T and T. Kawai 2009 *Appl. Phys. Lett.* **95** 261109
[6] Tomita Y, Suzuki N and Chikama K 2005 *Opt. Lett.* **30** 839
[7] Liu X, Adachi Y, Tomita Y, Oshima J, Matsubara K, Nakashima T and Kawai T 2012 *Opt. Express* **20** 13457
[8] Tomita Y, Hata E, Momose K, Takayama S, Liu S, Chikama K, Klepp J, Pruner C and Fally M 2016 *J. Mod. Opt.* **63** S1
[9] Dolgaleva K and Boyd R W 2012 *Adv. Opt. Photon.* **4** 1
[10] Sipe J E and Boyd R W 1992 *Phys. Rev. A* **46** 1614
[11] Liu X and Tomita Y 2012 *Phys. Res. Int.* **2012** 1
[12] Matsushima S and Tomita Y, to be presented at 2017 *the 24th Congress of the International Commission for Optics (ICO-24)* (Tokyo)
[13] Gu B, Fan Y, Wang J, Chen J, Ding J, Wang H and Guo B 2006 *Phys. Rev. A* **73** 065803
[14] Burda C, Link S, Green T C and El-Sayed M A 1999 *J. Phys. Chem. B* **103** 10775
[15] Bawendi M G, Wilson W L, Rothberg L, Carroll P J, Jedju T M, Steigerwald M L and Brus L E 1990 *Phys. Rev. Lett.* **65** 1623
[16] Liu X and Li Z 1995 *Solid-State Commun.* **96** 981
[17] Dolgaleva K, Shin H and Boyd R W 2009 *Phys. Rev. Lett.* **103** 113902
[18] Katayama K, Yamaguchi M and Sawada T 2003 *Appl. Phys. Lett.* **82** 2775
[19] Qu Y, Ji W, Zheng Y and Ying J Y 2007 *Appl. Phys. Lett.* **90** 133112

IOP Conf. Series: Journal of Physics: Conf. Series **867** (2017) 012012 doi:10.1088/1742-6596/867/1/012012

Nonlinear polarization rotation of two types of vector beams through isotropic Kerr nonlinearities

Bing Gu*, Bo Wen, Guanghao Rui, and Yiping Cui

Advanced Photonics Center, Southeast University, Nanjing 210096, China

*corresponding author: gubing@seu.edu.cn

Abstract. We report a theoretical investigation of the spatial self-phase modulation (SSPM) effect and nonlinear ellipse rotation of both hybridly polarized vector beams and elliptically polarized vector beams through isotropic Kerr nonlinearities. It is demonstrated that the SSPM intensity pattern, the distribution of state of polarization, and the spin angular momentum (SAM) flux of a hybridly polarized vector beam could be manipulated by tuning the magnitude of the isotropic optical nonlinearity. Interestingly, we observed the radial-variant nonlinear ellipse rotation of elliptically polarized vector beams through isotropic Kerr nonlinearities. Our results may find interesting applications in nonlinear mechanism analysis, nonlinear characterization technique, and SAM manipulation.

1. Introduction

The polarization-dependent optical nonlinearities that are associated with the tensor nature of the third-order susceptibility $\chi^{(3)}$ of optical materials are of direct relevance to technological applications in nonlinear photonics devices. One of the interesting effects is the nonlinear ellipse rotation that occurs when an intense elliptically polarized beam interacts with an isotropic medium owing to the existence of $\chi_{xyyx}^{(3)}$ [1]. The nonlinear polarization rotation effect has been exploited to discriminate the origin of optical nonlinearity [2], to characterize the third-order nonlinear susceptibility tensor [3], and in the nonlinear polarization switching [4].

Up to now, most of investigations on both the nonlinear polarization rotation and SSPM effects were excited by the scalar light field with homogeneously polarization. It is noteworthy that the cylindrical vector beam with inhomogeneously polarization has excited many novel nonlinear optical effects, such as second-harmonic generation [5], vectorial self-diffraction [6], and optical limiting [7]. Very recently, we have reported the polarization characteristics of radially polarized beams induced by an anisotropic Kerr nonlinearity [8].

In this work, we investigated the SSPM effect and nonlinear polarization rotation of both hybridly polarized vector beams and elliptically polarized vector beams through isotropic Kerr nonlinearities.

2. Theory and Results

Figure 1 shows the schematic for investigating the SSPM and nonlinear polarization rotation of two types of vector beams through isotropic Kerr nonlinearities. With the known incident field E_{in}, one can obtain the electric field E at the focal plane of a thin lens by the use of vectorial Rayleigh-Sommerfeld formulas. Using the thin sample approximation, the complex field at the exit plane of the sample E_e is obtained. Then the electric field distribution in the far-field observational plane E_a is obtained.

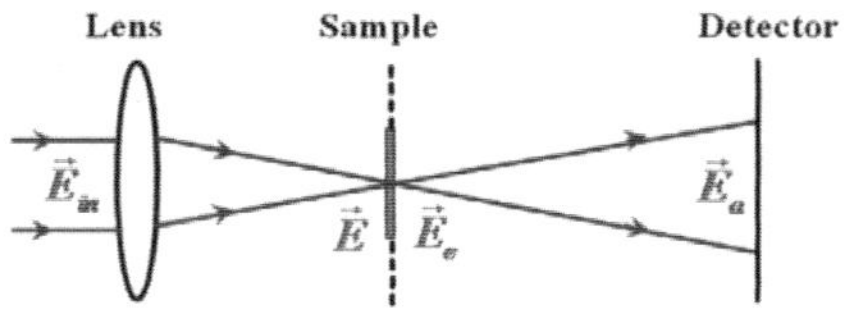

Figure 1. Schematic for investigating SSPM effect and nonlinear polarization rotation.

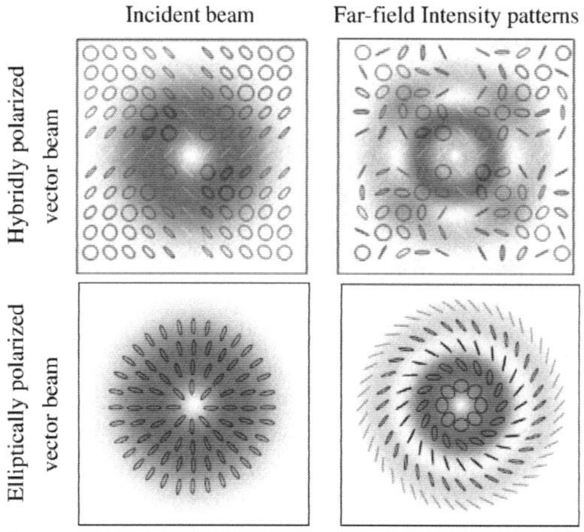

Figure 2. The incident and far-field intensity patterns with the SoP distributions of the hybridly and elliptically polarized vector beams, respectively.

We numerically simulate the far-field vectorial self-diffraction behaviors and nonlinear polarization rotations of two types of vector beams through isotropic Kerr nonlinearities. For instance, Fig. 2 illustrates the incident and far-field intensity patterns with the distributions of state of polarization (SoP) of the hybridly and elliptically polarized vector beams, respectively.

3. Conclusion

In summary, we systematically study the dependence of the self-diffraction behaviors on both the physical origin of the optical nonlinearity and the nonlinear phase shift. It is demonstrated that the SSPM intensity pattern, the SoP distribution, and the SAM flux of a hybridly polarized vector beam could be manipulated by tuning the magnitude of the isotropic optical nonlinearity. Interestingly, we observed the radial-variant nonlinear polarization rotation of elliptically polarized vector beams through isotropic Kerr nonlinearities. Our results may find interesting applications in nonlinear mechanism analysis, nonlinear characterization technique, and SAM manipulation.

Acknowledgements

This work was supported by the National Science Foundation of China (11474052, 11505049).

References

[1] Maker P D, Terhune R W, and Savage C W 1964 *Phys. Rev. Lett.* **12** 507
[2] Miguez M L, Barbano E C, et. al. 2014 *Opt. Express* **22** 25530
[3] Liu Z B, Yan X Q, et. al. 2007 *Opt. Express* **15** 13351
[4] Yin L, Zhang J, et. al. 2009 *Opt. Lett.* **34** 476
[5] Bautista G, Mäkitalo J, et. al. 2015 *Nano Lett.* **15** 1564
[6] Gu B, Ye F, et. al. 2012 *Opt. Express* **20** 149
[7] Wu J L, Gu B, et. al. 2014 *Appl. Phys. Lett.* **105** 171113
[8] Gu B, Wen B, et. al. 2016 *Opt. Lett.* **41** 1566

IOP Conf. Series: Journal of Physics: Conf. Series **867** (2017) 012013 doi:10.1088/1742-6596/867/1/012013

Ultrafast dynamics of photoexcited free carriers in CVD diamonds

Xiangming Liu, Ben Zhang, Quanjie Zhong, Xiaoshi Peng, and Shenye Liu*

Research Center of Laser Fusion, China Academy of Engineering Physics, Mianyang 621900, China

*lsye1029@163.com

Abstract. We report on the experimental studies of the free-carrier dynamics and nonlinear optical properties in both bulk polycrystalline and single-crystal chemical vapor deposition (CVD) diamonds. The dynamics of nonlinear refraction and nonlinear absorption is measured by use of a time-resolved femtosecond pump-probe technique under UV excitation. Nonlinear refraction dynamics indicates a positive Kerr effect. Slow and fast components in nonlinear absorption transmittance are separated and used to determine the free-carrier lifetimes.

1. Introduction

High-purity diamond crystals have received much attention [1,2] owing to their potential applications in semiconductor physics and optoelectronics. Due to their transparency over a broad wavelength range from the UV to IR and to their excellent thermal conductivity, diamonds have become one of the major window materials. However, in the study of inertial confinement fusion (ICF) the high-intensity drive laser and laser plasma-induced X-ray can modify the optical properties of the window materials composed of diamonds, which leads to the changes in the strength and phase of the probe beam in diagnostic apparatuses and therefore to the increased uncertainty in the experimental results. Therefore, it is of crucial importance to investigate the ultrafast dynamics of the free carriers and nonlinear optical properties in bulk diamond.

In this work, we investigate the free-carrier dynamics and nonlinear optical properties in both bulk polycrystalline and single-crystal chemical vapor deposition (CVD) diamonds by use of a time-resolved femtosecond pump-probe technique employing a phase object. The dynamics of nonlinear absorption and refraction is determined by the differential transmittance in open- and closed-aperture configuration excited by the UV laser, respectively.

2. Samples and experimental setup

Both bulk polycrystalline and single-crystal diamonds of high purity were studied in this paper. Diamonds were prepared by Element Six, UK employing chemical vapor deposition method. We used a modified pump-probe technique with a phase object (PO pump-probe technique) [3] to measure the dynamics of transmittance changes. A mode-locked Yb:KGW-based femtosecond laser (515 nm, 190 fs, 20 Hz) was divided into a pump and a probe beams. An optical parametric amplification (OPA) system was employed to change the pump beam to the UV light. A PO, which consisted of a glass plate with a transparent dielectric disc deposited on it, was introduced into the probe beam. The pump and probe beams were focused on the same point of the sample. After passing through the sample, the probe beam was divided into two beams. A finite aperture was placed in front of one detector and its radius in the far field was adjusted to be equal to that of diffracted beam of the PO. Energy change of the probe beam was recorded as a function of delay time between the pump and probe beam.

3. Experimental results

Figures 1(a)-1(c) show the normalized transmittances in open-aperture configuration as a function of delay time for polycrystalline diamond. Transmittance changes indicate that nonlinear absorption includes the non-degenerate two-photon absorption (ND-TPA) and free-carrier absorption [4]. The ND-TPA dominates near the zero delay and the free-carrier absorption decays slowly with a fast (80 ps) and a slow (3-8 ns) components. A sharp peak in the vicinity of zero time delay for a closed-aperture configuration in Fig. 1(d) can be ascribed to a positive nonlinear refraction due to the Kerr effect. The FWHM of the trace is 0.3 ps, which is associated with the pulse width. Similar results are obtained for other excitation densities. The nonlinear refractive index of the diamond is extracted from the fitting to be $(1.4\pm0.2)\times10^{-15}$ cm^2/W. Moreover, under high energy density the negative nonlinear refraction due to the free carriers will be observed at large time delay.

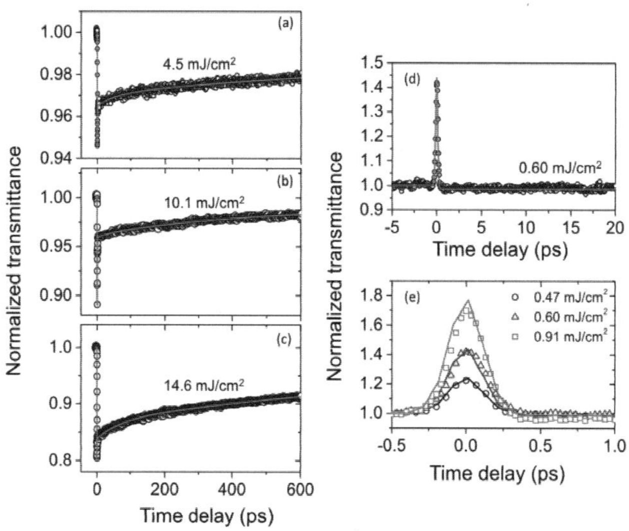

Fig. 1. Dynamics of (a)-(c) nonlinear absorption and (d)-(e) nonlinear refraction for the polycrystalline diamond under 351 nm excitation.

In addition, the transmittance changes in open-aperture configuration ranging from 480 to 680 nm are also investigated for both polycrystalline and single-crystal diamonds. The detailed results will be discussed somewhere else.

4. Conclusion

The optical nonlinearities and ultrafast free-carrier dynamics in both bulk polycrystalline and single-crystal CVD diamonds have been investigated by the PO pump-probe technique. We have found that both ND-TPA process and Kerr nonlinearity are positive. The free-carrier absorption (refraction) induced by the ND-TPA process is positive (negative). The decay of free-carrier absorption is relatively slow. This work will enable us to gain a better understanding of ultrafast dynamics of carriers and the nonlinearities in diamonds and therefore is beneficial to their applications in optics.

References

[1] Eremets M I *et al* 2005 Appl. Phys. Lett. **87** 141902
[2] Naka N *et al* 2008 Phys. Rev. B **80** 035201
[3] Yang J *et al* 2009 Opt. Express **17** 7110
[4] Zhang B *et al* 2016 Optik **130** 1073

IOP Conf. Series: Journal of Physics: Conf. Series **867** (2017) 012014 doi:10.1088/1742-6596/867/1/012014

Triboelectric Nanogenerator Using Lithium Niobate Thin Film

Juan Geng, Xinzheng Zhang, Yongfa Kong and Jingjun Xu
The MOE Key Laboratory of Weak-Light Nonlinear Photonics, TEDA Institute of Applied Physics and School of Physics, Nankai University, Tianjin 300457, P. R. China

E-mail: zxz@nankai.edu.cn

Abstract. We present a triboelectric nanogenerator (TENG) using a lithium niobate thin film, as one of the triboelectric pairs which was grown on a silicon substrate by laser molecule beam epitaxy (LMBE). The designed TENG has the advantages of simple structure, easy fabrication, small size (1.1*1.0*0.15 cm^3). An open-circuit voltage of 136 V and a short-circuit current of 8.40 μA have been achieved. The maximum output power is 307.5μW under the load resistance of 10MΩ. This is the first time to use lithium niobate thin film as one of the friction pair, which may make it possible to expand the application of triboelectric nanogenerator to optical field.

1. Introduction

As the aggravation of the environmental pollution and energy crisis, it is urgent to develop green energy sources. Since mechanical energy is very common and stable in our daily life, lots of research has been done to develop new methods and devices that can harvest mechanical energy. In 2012, Wang et al. invented triboelectric nanogenerator (TENG) [1] which was based on the conjunction of contact electrification and electrostatic induction. It can harvest many forms of mechanical energy, such as vibrational energy [2] and biomechanical energy [3]. TENG can generate electrical output when two materials with different friction polarities contact and separate from each other. Due to its low cost, easy fabrication, small size, lightweight and high power output, TENG has been proved to be a powerful means to convert mechanical energy into electricity.

It is crucial to develop rational structure to improve the output characteristic, especially to develop TENG with new fractional materials. Here, we demonstrate a new fractional material, lithium niobate thin film (LNTF). As a typical piezoelectric and pyroelectric material, lithium niobate is widely used in optical and material fields [4, 5]. Using LNTF as one of the tribo-pairs may make it possible to expand the application of TENG to optical or material field. Continuous periodic output is achieved when LNTF contacts and separates with micro-structured PDMS from black silicon mold. The resistive load characteristics of the TENG is also investigated.

2. Results and discussion

Working as the attached-electrode contact-mode, a TENG includes two layers as sketched in figure 1. A PDMS film with a thickness of 180 μm is selected to be one of the triboelectric layers because it is easy to gain electrons [6]. A black silicon [7] fabricated by a femtosecond laser is used as a mold to fabricate micro-structured PDMS. Served as an electrode, an aluminum foil is fixed on the PDMS by ordinary adhesive tape. The other layer is an iron doped LNTF which was grown on a highly doped silicon substrate by laser molecule beam epitaxy (LMBE) method. The silicon substrate serves as the other electrode to accumulate induced charges. The size of the TENG is 1.1*1.0*0.15 cm^3.

Figure 1. Structure of the triboelectric nanogenerator (TENG)

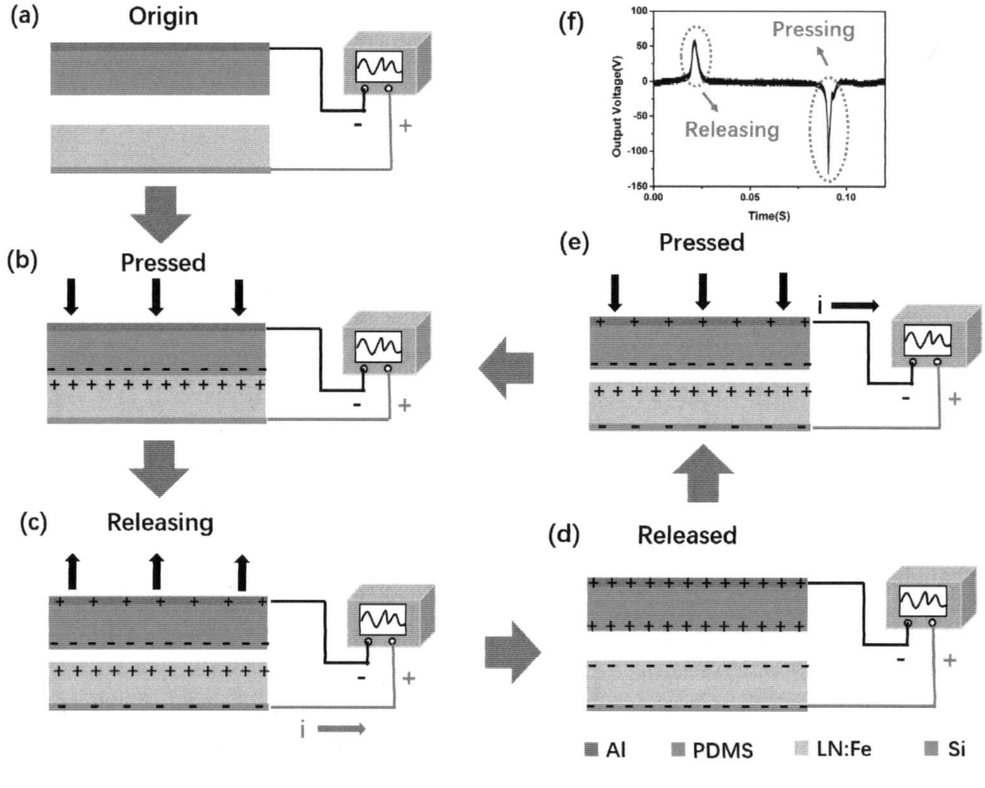

Figure 2. Working principle of the TENG using a LNTF. (a) original state with a fixed distance. (b) The friction pairs generate triboelectric charges when compressive force is applied. (c) Positive signal is generated while withdrawing the compressive force. (d) The system reaches electrical equilibrium. (e) Negative signal is generated when compressive force is applied again. (f) Open-circuit voltage during one cycle.

The TENG is based on the conjunction of contact electrification and electrostatic induction. In the beginning, the LNTF is separated from the PDMS film at a fixed distance. There are no triboelectric charges on the friction pairs and no charges transfer between the electrodes, which means no current flow (Figure 2(a)). By applying compressive force with a vibration platform, the LNTF starts to contact with the PDMS film. Because of the contact electrification, the electrons are transferred from the LNTF to the PDMS film since it is much easier for PDMS to gain electrons. As a result, the opposite triboelectric charges with the same density are generated on the inner surfaces of the LNTF and the PDMS film (Figure 2(b)). As the compressive force is withdrawn, the LNTF starts to separate from the PDMS film, leading to a potential difference between them. In order to screen this potential difference, the electrons will flow from the Al electrode to the Si electrode (Figure 2(c)) generating a positive signal in the external circuit (Figure 2(f)). When the system reaches electrical equilibrium, there will be no signal. (Figure 2(d)). When the compressive force is applied again, it will build a reversed potential difference, which will drive the inductive electrons to flow in the opposite direction (Figure 2(e)), and thus generates a negative signal in the external circuit (Figure 2(f)). Then, the LNTF and the PDMS film are in contact again with no signal output (Figure 2(b)). This is one cycle of the electricity generation process.

The output measurement is investigated with a separation distance of 1.5 mm and a vibration frequency of 9.0 Hz. The periodic compressive force with variable frequency and amplitude is achieved by a connection of a waveform generator (Agilent 33120A), a power amplifier (SINOCERA YE5872A), and a vibration platform (SINOCERA JZK-10). The output voltage and current are measured by an oscilloscope (Tektronix DPO4054). Figure 3(a) demonstrates an open-circuit voltage of 136 V measured by a high voltage differential probe (Tektronix THDP0100). Figure 3(b) shows a short-circuit current of 8.4 μA measured with a low value resistor of 24 kΩ. Both the voltage and current signals are very steady, which indicates the stability of the TENG.

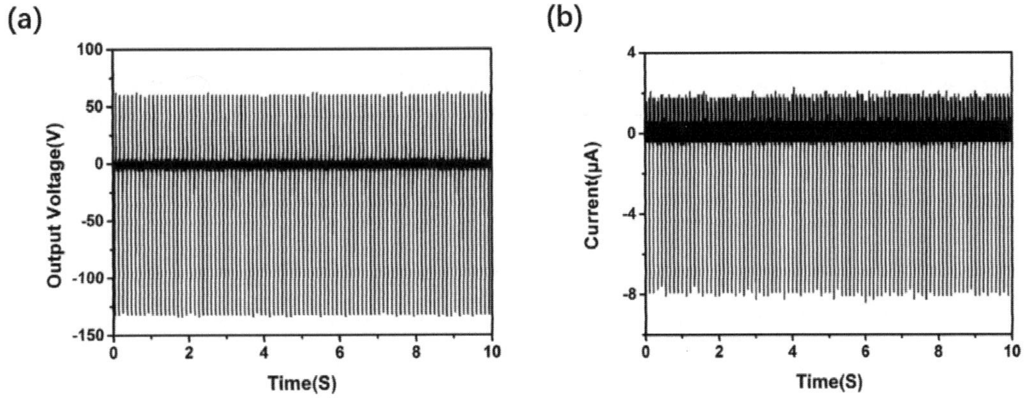

Figure 3. output characteristics. (a) open-circuit voltage, (b) short-circuit current.

As for practical applications, the electrical output to external devices varies a lot due to the matching between the external load and the TENG. In order to investigate the optimum resistive load, we have measured the voltage output under different external resistive load from 1.00 MΩ to 100 MΩ, as sketched in figure 4(a). When the resistance is very small, the influence to the real charge transfer is not obvious, the external current is nearly short-circuit current. As the resistor increases, it starts to affect the charge transfer rate dramatically, resulting in a rapid drop of the current. When the resistance is large enough, there will be nearly no charge transfer between two electrodes, and the voltage will approach to open-circuit voltage. The dependence of output power curves on the resistance is shown in figure 4(b). The power reaches its maximum value when the external resistance is equal to the internal resistance of the TENG. With a load resistance of 10 MΩ, the TENG gives the maximum power output of 307.5 μW, corresponding to a power density of 280 μW/cm^2.

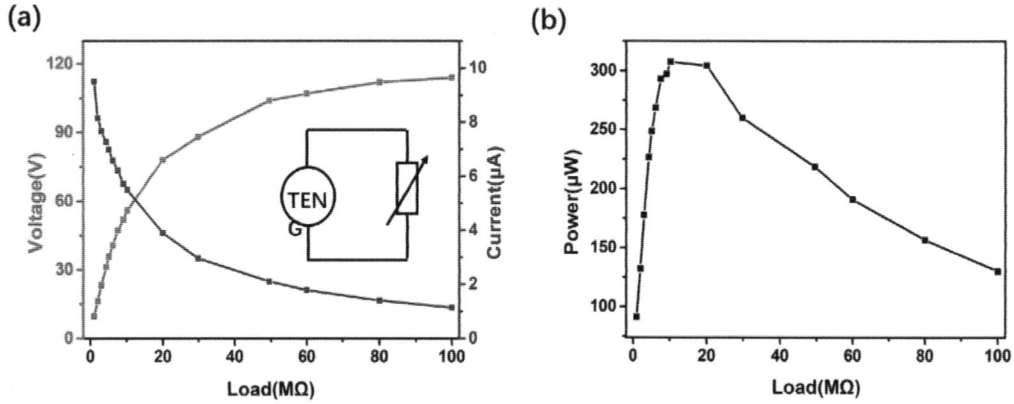

Figure 4. Resistive load characteristics. (a) peak voltage and current as a function of external load resistance. (b) instantaneous power output as a function of external load resistance.

3. Conclusions

In summary, we present a triboelectric nanogenerator based on a brand new fraction material, LNTF. This device is stable, easily fabricated and low cost. The volume of the device is 1.1*1.0*0.15cm^3. It can provide an open-circuit voltage of 136V and a short circuit current of 8.4μA. When the load resistance is 10 MΩ, the TENG can give a maximum power output of 307.5 μW. This work shows the possibility of using lithium niobate as a new friction material of TENG to convert mechanical energy into electricity energy, which may expand the application fields of TENG.

References

[1] Fan F R, Tian Z Q and Wang Z L 2012 *Nano Lett.* **1** 328
[2] Zhang X S, Han M D, Wang R X, Zhu F Y, Li Z H, Wang W and Zhang H X 2013 *Nano Lett.* **13** 1168
[3] Bai P, Zhu G, Lin Z H, Jing Q S, Chen J, Zhang G, Ma J S and Wang Z L 2013 *ACS Nano* **7** 3713
[4] Dunn M H and Ebrahimzadeh M 1999 *Science* **286** 1513
[5] Ilchenko V S, Savchenkov A A and Matsko A B 2004 *Phys. Rev. Lett* **92** 043903
[6] Wang Z L 2013 *ACS Nano* **7** 9533
[7] Her T H, Finlay R J, Wu C, Deliwala and Mazur E 1998 *Appl. Phys. Lett.* **73** 1673

Photorefractive trimming of Ti in-diffused waveguides for high performance lithium niobate photonic integrated circuits

A. Tronev[1,2], M. Parfenov[3], D. Dolbik[3], P. Agruzov[1], I. Ilichev[1] and A. Shamray[1,2,3]

[1] Ioffe Institute, Polytekhnicheskaya 26, 194021 St.Petersburg, Russia
[2] ITMO University, Kronverksky av. 49, 197101 St.Petersburg, Russia
[3] Peter the Great Saint-Petersburg Polytechnic University, Polytekhnicheskaya 29, 195251 St.Petersburg, Russia
e-mail: achamrai@mail.ioffe.ru

Abstract. A local micro-size change of refractive index due to photorefractive effect was effectively used for precise adjustment of the splitting ratio of waveguide optical splitters. The optimal conditions for efficient control of a splitting ratio for two basic configurations (X-coupler and Y-branch) were found. To demonstrate the efficiency of the proposed method the adjustment of extinction ratio of Mach-Zehnder integrated optical modulator was performed. The increase of 17 dB (from 30 to 47 dB) was obtained. The results are very promising for fabrication high extinction ratio modulators for quantum key distribution, precise sensing, high fidelity signal processing and others.

1. Introduction

Lithium niobate (LN) is one of basic materials of integrated optics used, for example, for electro and acousto-optical modulators, switches, diffraction gratings, nonlinear optical wavelength converters, and others [1]. LN-based ion exchange or Ti-indiffused photolithographic techniques are established technologies, however, some modern applications, such as high-purity suppressed carrier synthesis or high extinction ratio pulse generation for precision measurement systems require the use of additional active trimming electrodes [2] and a corresponding servo loop because of finite tolerances of optical elements. Recently developed direct laser writing (DLW) with femtosecond/ultrafast laser pulses [3] can be employed to address the problem, but this method is problematic for mass production and suffer from high insertion loss in optical waveguides (typically ~ 1 dB/cm compared to <0.1 dB/cm for Ti-indiffused waveguides). A method that combines low loss of Ti in-diffused technology and passive permanent trimming of DLW was proposed. We demonstrated that a local micro-size change of refractive index can be effectively used for precise adjustment of the splitting ratio of waveguide optical splitters and couplers and does not impairs the optical losses.

2. Experimental results

The experimental setup for a photorefractive trimming is shown on the Fig. 1. A He-Ne laser (633 nm) illumination was used for the local refractive index change. It was directed by standard single-mode optical fiber with the tip fixed close to the surface of an integrated optical chip. A rather lower light power (4.5 mW) at the fiber output was applied. The precise positioning of the light spot on an integrated optical scheme was performed by the displacement of the chip which was fixed on the 3 axis translation stage. The trimming was monitored by test fiber optic circuit at the wavelength 1550 nm.

The trimming of optical waveguide splitters of two types (X direction coupler and Y branch) was investigated. The dependence of splitting ratio change on position of the light spot relative waveguide topology was studied (Fig. 2, 3). The position corresponding to the maximum change depended on the light distribution between coupled waveguides in the X coupler. It was in the center between waveguides for the parallel section where light distribution was symmetrical and shifted to the side of higher intensity at the input approaching and output separation sections. For the Y branch the influence on the different branches lead to the opposite changes of a splitting ratio which was caused initial asymmetry due to technological inaccuracy. A typical time for the reaching a steady-state condition was about 2 minutes and agreed with a photorefractive response time. A homogeneous illumination of the chip by intense visual or UV light erased a local refractive index change and returned splitters to the initial states. The thermal fixing [4] or intense femtosecond laser writing [1] could be used for an undecayable trimming.

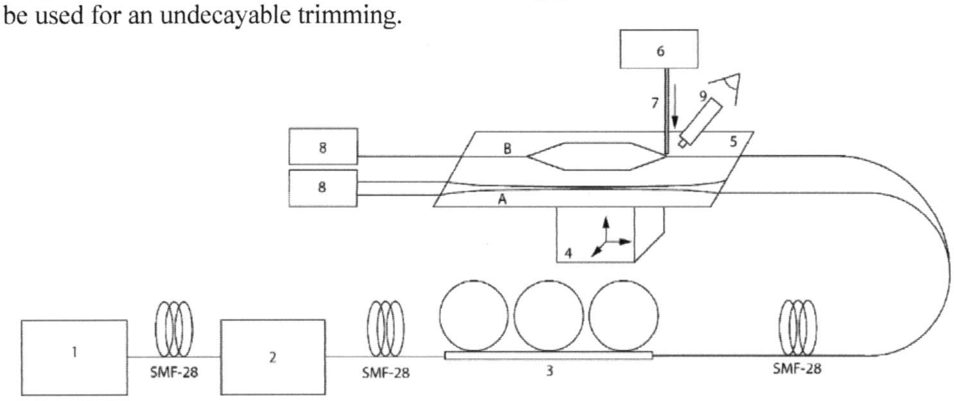

Figure 1. Experimental setup. 1 – DFB-laser diode, 2 – fiber optical attenuator, 3 – polarization controller, 4 – 3-axis translation stage, 5 – integrated optical chip, 6 – He-Ne laser, 7 – single mode optical fiber, 8 – optical detector, 9 – microscope.

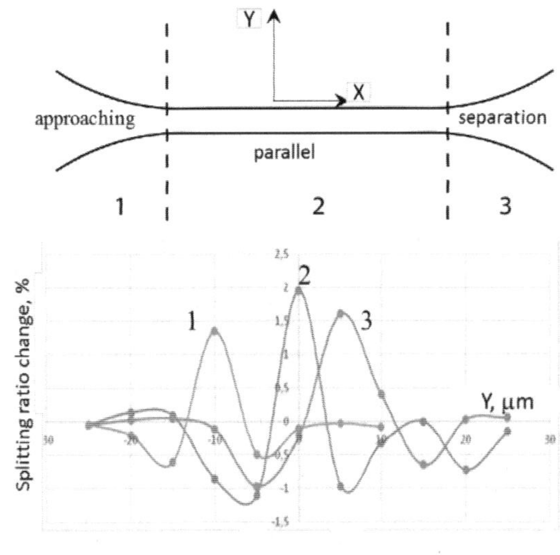

Figure 2. Trimming of the X coupler: splitting ratio change versus light spot position.

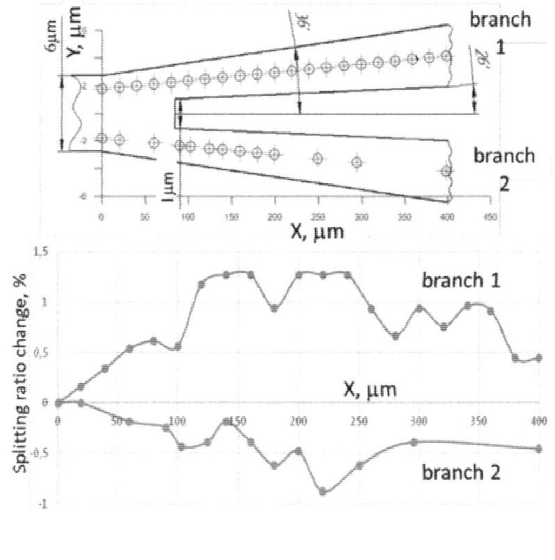

Figure 3. Trimming of the Y branch: splitting ratio change versus light spot position.

3. Conclusion

In conclusion note that rather small changes of a splitting ratio (1 – 2 %) demonstrated in the experiments can significantly influence on integrated optical device performance. To demonstrate the efficiency of the proposed method the adjustment of extinction ratio ER (max/min ratio of transmitted

IOP Conf. Series: Journal of Physics: Conf. Series **867** (2017) 012015 doi:10.1088/1742-6596/867/1/012015

optical power) of Mach-Zehnder integrated optical modulator was performed. The increase of 17 dB (from 30 to 47 dB) was obtained. The results are very promising for fabrication high extinction ratio modulators for quantum key distribution, precise sensing, high fidelity signal processing and others.

References

[1] Bazzan M and Sada C 2015 *Appl. Phys. Rev.* **2** 040603
[2] Yamaguchi V Y, et al 2014 *Japan. J. Appl. Phys.* **53** 08MB03
[3] Chen F and Aldana J R V 2013 *Laser Photonics Rev.* 1
[4] Grachev A I, Chamrai A V and Petrov M P, 2001 *OSA TOPS* **62** 203

IOP Conf. Series: Journal of Physics: Conf. Series **867** (2017) 012016 doi:10.1088/1742-6596/867/1/012016

Sum Frequency and Second Harmonic Generations in Lithium Niobate Microdisk Resonators on a Chip

Zhenzhong Hao,[1] Jie Wang,[1] Shuqiong Ma,[1] Fang Bo,[1,*] Feng Gao,[1] Guoquan Zhang,[1,2] Jingjun Xu,[1,3]

[1]*The MOE Key Laboratory of Weak Light Nonlinear Photonics, TEDA Institute of Applied Physics and School of Physics, Nankai University, Tianjin 300457, China*

Email: [*] bofang@nankai.edu.cn, [2]zhanggq@nankai.edu.cn, [3]jjxu@nankai.edu.cn

Abstract: We report both sum frequency and second harmonic generations in on-chip lithium niobate microdisk resonators under the pump of two individual lasers. The conversion efficiency of sum frequency generation was measured to be 8.3×10^{-9} mW^{-1}, which are on the same order of the accompanied second harmonics. The dependence of upconversion signals on the wavelength of one pump laser was investigated.

1. Introduction

Lithium niobate whispering gallery mode (WGM) microresonators are a good platform for nonlinear frequency conversion due to outstanding nonlinear optical properties of lithium niobate crystal and the enhanced light-material interactions in the microresonators. Various nonlinear optical effects were observed in millimeter sized lithium niobate WGM resonators with quality factors of the order of 10^7. Recently, on-chip lithium niobate microdisk resonators with quality factors higher than one million was fabricated from lithium niobate on insulator film [1]. Second harmonic generation (SHG) [2-4] was reported in such on-chip microdisk by several groups. Here we report the first observation of sum frequency generation (SFG) accompanied by SHGs in lithium niobate microdisk resonators on a chip.

2. Experimental setup for nonlinear optical effects investigation

Light from two individual lasers in 1550 nm band functioning as pumps were launched into a lithium niobate disk with quality factor of the order of 10^5 by using a tapered fiber. The tapered fiber was also used to collect SFG and SHG signals. These signals were then analyzed utilizing a spectrometer. The wavelength and output power of each pump laser can be tuned separately. Therefore, the dependence of the nonlinear signals on experimental parameters was studied systematically.

3. SFG and SHG in lithium niobate microdisks

We first sent two pump lasers into the lithium niobate microdisk simultaneously. By adjusting the wavelength of the two pumps, we observed sum frequency generation signal from the grating spectrometer. We also observed second harmonic generation signals of two two pumps when both of the input wavelengths and the signal wavelengths matched the resonances of the cavity. Sum frequency generation together with the second harmonic generation signals are shown in different colors in Fig.1. The conversion efficiency of the SFG signal was measured to be 8.3×10^{-9} mW^{-1}.

We also investigated the variation of the intensities of the SFG and SHG signals with respect to the frequency detuning of one pump laser about the resonance of the lithium niobate microdisk. The wavelength of the other laser was fixed during the measurement. The measured results are shown in Figs. 1(b-d). It is seen from those figures that the SFG and SHG of the laser whose wavelength was tuned reach their maximums twice when the pump wavelength passed through two nearby resonances of the lithium niobate resonator. On the contrast, the SHG of the laser with fixed wavelength obtained minimum signal intensities due to energy consumption of the SFG process.

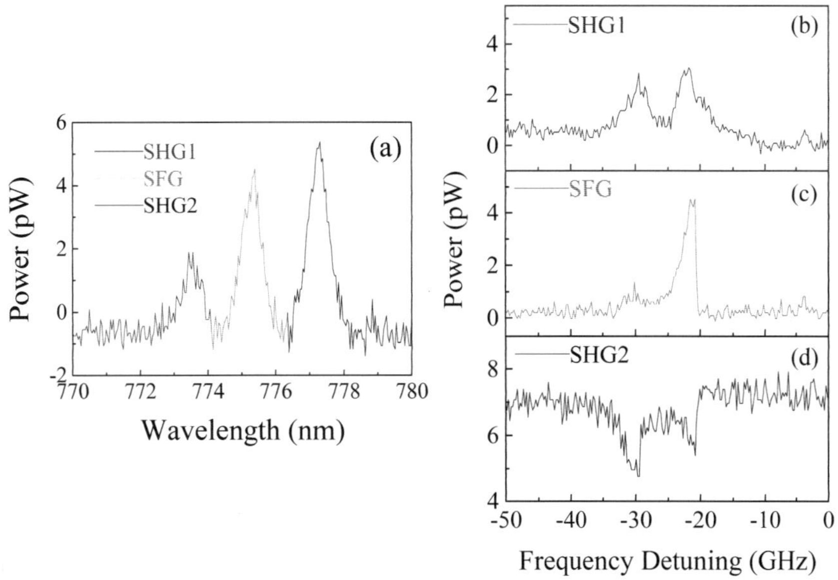

Fig. 1. (a) Spectra of nonlinear optical signals. SFG (red line) and SHG (blue and black lines) signals obtained when the wavelengths of the two pump laser were set to be 1546.9 nm and 1554.6 nm, respectively. (b-d) Dependences of nonlinear optical signal intensities on the wavelength of one pump laser measured when the wavelength of the other laser was fixed.

4. Conclusions

In conclusion, SFG signal together with SHGs were observed and studied systematically in lithium niobate microdisk resonators on a chip. The SFG process in on-chip lithium niobate WGM resonators may be used in applications ranging from infrared single photon detection to sum-frequency vibrational spectroscopy.

Acknowledgment

Z. Hao and J. Wang contributed equally to this work. F. Bo, F. Gao, G. Zhang, and J. Xu are also at the Collaborative Innovation Center of Extreme Optics, Shanxi University, Taiyuan, Shanxi 030006, China. This work was financially supported by 973 program (2013CB328702); Natural National Science Foundation of China (NSFC) (11674181, 11374165, 11674184, and 61475077); 111 Project (B07013); PCSIRT (IRT_13R29).

References

[1]. Wang J, Bo F, Wan S, Li W, Gao F, Li J, Zhang G and Xu J 2015 *Opt. Express* **23**(18) 23072-23078

[2]. Wang C, Burek M J, Lin Z, Atikian H A, Venkataraman V, Huang I, Stark P and Loncar M 2014 *Opt. Express* **22** 30924-30933
[3]. Lin J, Ying X, Fang Z, Min W, Wang N, Qiao L, Wei F and Cheng Y 2015 *Sci. China Phys. Mech.* **58** 114209
[4]. Lin J, Ying X, Ni J, Wang M, Fang Z, Qiao L, Wei F and Cheng Y 2016 *Phys. Rev. Appl.* **6** 014002

Linear diffraction of light waves on periodically poled domain structures in lithium niobate crystals: collinear, isotropic, and anisotropic geometries

S M Shandarov[1], A E Mandel[1], T M Akylbaev[1], M V Borodin[1], E N Savchenkov[1], S V Smirnov[1], A R Akhmatkhanov[2], and V Ya Shur[2]

[1]Department of Electronic Devices, Tomsk State University of Control Systems and Radioelectronics, 40, Lenin Ave., Tomsk 634050, Russia
[2]Institute of Natural Sciences, Ural Federal University, 51, Lenin Ave., Ekaterinburg 620083, Russia

E-mail: stanislavshandarov@gmail.com

Abstract. The possible variants of experimental observation of light diffraction on periodically poled domain structures (PPDS) in the lithium niobate crystal with 180-degree domain Y-walls are considered. We experimentally investigated isotropic and anisotropic diffraction of coherent light (λ = 655nm) on the PPDS with spatial period Λ = 8.79 μm produced by poling method in a $LiNbO_3$: 5% MgO crystal. The central wavelength of irradiation experiencing a collinear diffraction on these PPDS is estimated as λ_c = 455 nm.

1. Introduction

Linear light diffraction on periodically polarized domain structures (PPDS) in ferroelectric crystals is an effective method of nondestructive testing of their characteristics and quality [1, 2]. In the absence of external applied electric field, such diffraction is caused by perturbations of the optical properties by crystal domain walls [1, 3, 4]. Within the framework of the known model of polarization distribution over the domain wall [5], expressions for the perturbations of the dielectric impenetrability components created by it in crystals with $3m$ symmetry have been obtained in [3]. From them it follows the possibility of realization of collinear and anisotropic diffraction on the PPDS in addition to isotropic diffraction. Experimentally, collinear diffraction was observed in a $LiNbO_3$:5 % MgO crystal on the PPDS with the spatial period Λ = 6.89 μm [3]. Authors [4] have been studied theoretically and experimentally the isotropic process with many diffraction maxima realized when the incident laser beam propagated along the Z axis of the $LiNbO_3$:5 % MgO crystal with the PPDS having the spatial period Λ = 9.43 μm.

In the present work, the special features of the isotropic and anisotropic Bragg diffractions of the coherent light beam propagating in the XY plane of the lithium niobate crystal in which PPDS with Y-walls having the spatial period Λ = 8.79 μm were produced by the polarization reversal method, are considered.

2. Materials

The examined PPDS were manufactured by Labfer LTD in the $LiNbO_3$: 5% MgO crystal having sizes 40 × 2 × 1 mm^3 along the X, Y, and Z axes, respectively, by the polarization reversal method in an external spatially-periodic electric field [6]. The periodic structure with $\Lambda = 8.79$ μm occupied completely the space between the sample faces perpendicular to the Z (h = 1 mm), Y (d = 2 mm) and X (L = 40 mm) axes.

3. Experimental results

Results of experiments on measuring the efficiency of the Bragg diffraction of ordinary (the subscript "o") or extraordinary (the subscript "e") light beam with the wavelength $\lambda = 655$ nm on the components of the Fourier spectrum of perturbations of optical properties with spatial frequencies $2\pi p/\Lambda$ (p = 1, 2, 3, ...) into the diffracted beam with corresponding polarization are presented in Table 1.

Table1. Efficiency of the Bragg diffraction of light beams propagating in the XY plane

p	1	2	3	4	5	6	7	8
$\eta_{oo}\times100$	0.219	0.526	0.113	0.028	0.045	0.010	0.003	0.002
$\eta_{ee}\times100$	0.743	7.170	1.273	0.447	0.513	0.064	0.063	0.022
$\eta_{oe}\times100$	–	–	–	0.102	0.023	0.025	0.006	0.002
$\eta_{eo}\times100$	–	–	–	0.130	0.022	0.030	0.017	0.004

As follows from Table 1, the intensity of the odd maxima of the isotropic Bragg diffraction is nonzero, and its peak efficiency is observed for the second orders ($p = \pm2$). These specific features can be caused by the fact that two periodic systems of walls produced during synthesis of the PPDS with polarization changing along the x coordinate from $-\boldsymbol{P}_S$ to $\boldsymbol{P}_S$ and from $\boldsymbol{P}_S$ to $-\boldsymbol{P}_S$ can have the spatial shift that differs from the half-period $\Lambda/2$ by a certain value Δx [4].

A comparison of the theoretical dependence of the anisotropic diffraction efficiency on its order p with the experimental data presented in Table 1 demonstrated that the main contribution comes from perturbations of the components $\Delta\varepsilon_{23} = \Delta\varepsilon_{32}$ of the dielectric permittivity tensor in the PPDS caused by the electrostriction and photoelasticity effects [3]. This makes it possible to realize the collinear diffraction on the given PPDS of the light beam with the wavelength $\lambda_c = 455$ nm propagating along the X axis of the crystal.

4. Conclusions

The isotropic and anisotropic diffractions of coherent light beams on the PPDS with the spatial period $\Lambda = 8.79$ μm produced in the $LiNbO_3$: 5% MgO crystal have been studied experimentally. We have also demonstrated the possibility of collinear diffraction of light with a wavelength of 455 nm on these PPDS.

Acknowledgments

This work was supported by the State Assignment of the Ministry of Education and Science of Russian Federation for 2017–2019 (No. 3.8898.2017/8.9, 3.4993.2017/6.7, 3.4973.2017/7.8), the RFBR (grant No. 16-29-14046-ofi_m) and the Government of the Russian Federation (Act 211, Agreement 02.A03.21.0006). The equipment of the Ural Center for Shared Use "Modern nanotechnology" UrFU was used.

References

[1] Aleksandrovskii A L, Gliko O A, Naumova I I and Pryalkin V I 1996 *Quant. Electron.* **27** 641
[2] Müller M, Soergel E, Buse K, Langrock C and Fejer M M 2005 *J. Appl. Phys* **97** 044102
[3] Shandarov S M, Mandel A E, Smirnov S V, Akylbaev T M, Borodin M V, Akhmatkhanov A R, and Shur V Ya 2016 *Ferroelectrics*, **496** 134
[4] Shandarov S M, Mandel A E, Andrianova A V, Bolshanin G I, Borodin M V, Kim A Yu, Smirnov S V, Akhmatkhanov A R, and Shur V Ya 2017 *Ferroelectrics*, **508** 49
[5] Zhirnov V A *Soviet Physics Jetp-USSR* 1959 **8** 822
[6] Shur V Ya *Ferroelectrics* 2008 **340** 3

Fabrication of nano-scale domain structures in bulk Mg-doped $LiNbO_3$ crystals

Yuejian Jiao, Xiaojie Wang, Yunfan Zhang, Fang Bo, Yongfa Kong, Jingjun Xu,and Guoquan Zhang*

The MOE Key Laboratory of Weak Light Nonlinear Photonics, School of Physics and TEDA Applied Physics Institute, Nankai University, Tianjin 300457, China

zhanggq@nankai.edu.cn

Abstract. We present a new method to fabricate nano-scale domain structures in bulk Mg-doped $LiNbO_3$ crystals with the assistance of thermal fixing technique. Domain structures with hundred-nanometer width and millimeter length can be fabricated using this method. SEM image reveals a feature of smooth sidewall for the nano-scale domain structures.

1. Introduction

Domain nanotechnology, a new branch of technology for the fabrication of micro/nano-scale domain structures with good reproducibility, has become one of the most important subjects of ferroelectric science and technology [1,2]. Benefit to the excellent electro-optic, piezoelectric, pyroelectric and nonlinear optical properties, $LiNbO_3$ has attracted much attention, especially in the nano-domain engineering, which allows for the generation of counter propagating photons in nonlinear optical applications [3]. The common method of domain engineering in $LiNbO_3$ is direct electric field poling [4-7]. Unfortunately, this traditional technique shows difficulty in fabricating sub-micro scale domain structures due to the domain broadening outside of electrodes [8].

Various techniques were developed to fabricate micro/nano-scale domain structures, for example, pulse laser irradiation [9], backswitch poling [10], multiple-pulse electric field poling [11], electron beam irradiation [12], or direct domain inversion by AFM in $LiNbO_3$ thin film [13], but it is still difficult to fabricate high quality and stable nano-scale domain structures in bulk $LiNbO_3$ crystals. Here, we present a new method to fabricate nano-scale domain structures in bulk $LiNbO_3$ crystals.

2. Fabrication of nano-scale domain structures

The samples used in our work are z-cut heavily Mg-doped (5 mol%) congruent $LiNbO_3$ crystal wafers with a thickness of 500 μm. First, periodically poled lithium niobate (PPLN) crystals were fabricated by the traditional electric field poling technique. After that the PPLN was heated to a temperature above 80 °C to thermally activate the movement of protons in $LiNbO_3$ crystal to compensate the electric charges in the domain wall. The protons would be fixed in the domain wall which may induce pinning defects when the crystal temperature was below 80 °C. In the end, the PPLN was forward polled or backward polled using the electric field poling technique. Due to the proton-induced pinning detects, an increase of the coercive field in the domain wall will inhibit the domain inversion process in the secondary poling process, and a narrow domain structure with almost the same dimension with the domain wall would not be inversed in the secondary poling process. The nano-scale domain structures can be revealed by the chemical HF acid etching and then observed by the SEM.

Using this technique, a nano-scale domain structure with a width of 480 nm and a length of ~1 mm on the +Z surface of $LiNbO_3$ crystal can be fabricated. The SEM image of part of the nano-scale domain structure is shown in figure 1. From the SEM image we can see that the domain structure is of high uniformity in the longitudinal dimension. And the sidewall of the domain structure is straight and smooth.

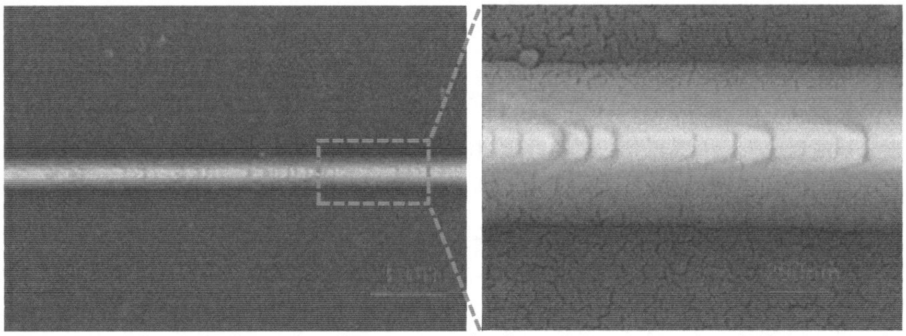

Figure 1. The SEM images of a nano-scale domain structure in $LiNbO_3$ crystal.

3. Conclusion

In conclusion, we proposed a new method to fabricate nano-scale domain structures in bulk $LiNbO_3$ crystal. Domain structure with a 480 nm width and ~1 mm length was fabricated. The decrease in the size of domain structures allows for large integration density. Our work may propel the applications of $LiNbO_3$ in nano-waveguides, nonlinear nano-optics and integrated nano-optics.

References

[1] Shur V Y 2008 *Ferroelectrics* **373.1** 1-10
[2] Shur V Y 2008 *Hand Book of Advanced Dielectric Piezoelectric and Ferroelectric Materials—Synthesis Characterization& Applications,* ed Ye Z G, (Woodhead)
[3] Canalias C, and Valdas P 2007 *Nature Photonics* **1** 459-462
[4] Byer R L 1997 *Journal of Nonlinear Optical Physics & Materials* **6** 549-592
[5] Yamada M, Nada N, Saitoh M, and Watanabe K 1993 *Applied Physics Letters* **62** 435-436
[6] Myers L E, Eckhardt R C, Fejer M M, Byer R L, Bosenberg W R, and Pierce J W 1995 *Journal of the Optical Society of America* B **12** 2102-2116
[7] Ross G W, Pollnau M, Smith P G R, Clarkson W A, Britton P E, and Hanna D C 1998 *Optics Letters* **23** 171-173
[8] Ye Z G 2008 *Handbook of Advanced Dielectric, Piezoelectric Ferroelectric Materials* (Woodhead)
[9] Shur V Y, and Zelenovskiy P S 2014 *Journal of Applied Physics* **116** 066802
[10] Shur V Y, Rumyantsev E V N E L, Shishkin E I, Fursov D V, Batchko R G, Eyres L A, Fejer M M, Byer R L 2000 *Applied Physics Letters* **76** 143-145
[11] Takashi S, and Toshiaki S 2013 *Japanese Journal of Applied Physics* **52** 100204
[12] Shur V Y, Chezganov D S, Smirnov M M, Alikin D O, Neradovskiy M M, and Kuznetsov D K 2014 *Applied Physics Letters* **105** 052908
[13] Gainutdinov R V, Volk T R, and Zhang H H 2015 *Applied Physics Letters* **107** 162903

Orientation and patterning of zeolite micro-crystals on photorefractive templates

I Elvira[1,2], J F Muñoz-Martinez[1,3], J Ramiro[3], A Barroso[2], C Denz[2], A García-Cabañes[1], F Agulló-López[1] and M Carrascosa[1]

[1]Dep. de Física de Materiales, Universidad Autónoma de Madrid, Madrid 28049, Spain
[2]Institut für Angewandte Physik, Westfälische Whihelms-Universität, Münster 48149, Germany
[3]Dep. Mecánica de Fluidos y Propulsión Aeroespacial, Universidad Politécnica de Madrid, Madrid 28040, Spain

Abstract: Evanescent fields photo-generated on the surface of photorefractive Fe-doped $LiNbO_3$ templates have been used to trap and align anisotropic zeolite micro-cylinders. *x*- and *z*-cut crystal configurations have been employed to obtain different 1D and 2D zeolite micro-patterns. Inside of the patterns, a substructure of zeolites aligned along the electric field, either parallel or perpendicular to the surface, have been obtained. The experiments have been analyzed to the light of a theoretical analysis that satisfactorily explains the main experimental features.

m.carrascosa@uam.es

1. Introduction

The photovoltaic tweezers (PVT) technique [1] takes advantage of the photorefractive (PR) effect presented in some ferroelectric crystals to efficiently trap and arrange different kind of objects. However, alignment of anisotropic objects with PVT has not been investigated in deep so far. Only a previous work mentioning orientation of living rod-shaped bacteria type *Escherichia coli* by photovoltaic (PV) fields has been recently reported [2].

In this communication, we demonstrate for the first time the versatile capabilities of PVT for massive alignment of anisotropic material on top of PR substrate. To this end, we have chosen zeolite L (ZL) micro-cylinders, a very interesting material with many technological applications [3], some of which require their alignment [4]. We study the role of the crystal configuration and the illumination pattern for obtaining different structures. The theoretical analysis agrees with the experimental results demonstrating that organization of these neutral objects is governed by the dielectrophoretic (DEP) potential whereas ZLs align parallel the PV electric fields.

2. Experimental procedure

ZL Linde Type cylinders of 4 μm length and 1.5 μm diameter have been used due to their high marked anisotropy. In order to fabricate ZL pattern on top of the PR substrate surface, the crystal is illuminated with an appropriate light pattern (at λ = 532 nm, with intensities in the range 1-100 mW/cm^2 and irradiation times around few minutes). After the photo-generation of PR fields, ZLs are deposited on the substrate by immersing the crystal in a suspension of ZLs in hexane [1]. The two different crystal configurations of $LiNbO_3$:Fe used are: *x*-cut (or *y*-cut) that contains the polar axis (*c*-axis) on the surface, and *z*-cut, with the *c*-axis perpendicular to the surface [5].

3. Results and discussion

The *x*-cut substrate has been illuminated with a fringe light pattern (periodicity Λ=18 μm) generated by the interference of two laser beams. Figure 1a displays the obtained particle pattern, showing that ZLs are trapped by PVT following the light periodicity. Inside of each fringe, one can also appreciate that ZLs are aligned horizontally, parallel to the *c*-axis. Simulations of the evanescent PR fields and DEP potential for this configuration have been carried out using a previous reported model [6]. They reveal the existence of a dominant electric field component along the *c*-axis and DEP potential minima with same periodicity of the light, consistent with the ZL alignment and spatial distribution, respectively.

According to simulations [7], *z*-cut experiments should allow vertical alignment of ZLs. Figure 1b shows an illustrative example of a *z*-cut sample that was illuminated using a triangle image (800 μm side) generated using a spatial light modulator. As expected, a high percentage of ZLs is placed vertically in the central area of the triangle (see inset in the figure). However, ZLs trapped in the boundaries of the triangle remain parallel to the surface and radially aligned with regard to the center of the triangle. The corresponding *z*-cut simulation is shown in Figure 1c. It explains satisfactorily the experimental results since the electric field is essentially vertical in almost the whole illuminated area, except at the triangle boundaries where it becomes parallel to the surface.

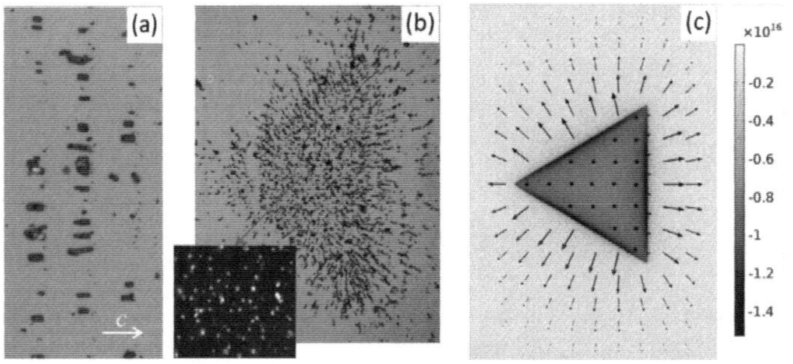

Figure 1: (a) ZL fringe pattern obtained by illuminating a *x*–cut sample with a sinusoidal pattern with Λ=18 μm periodicity, (b) ZL triangle pattern obtained in a *z*-cut sample illuminating with a triangle light image (800 μm side). (c) Simulation of the evanescent PV field (arrows) and DEP potential (colour degrade) for the experiment of the triangle. In both cases, ZLs are aligned along the evanescent electric field lines.

4. Summary and Outlook

The most relevant results of this novel application of PVT are the high flexibility to create different structures of aligned ZLs and the good agreement between theory and experiments. Massive trapping and alignment of ZLs have been reached with low light intensity and in a few minutes. These aligned structures should be very appropriate for further applications such as enhancement of non-linear properties or energy transfer processes, to name just a few.

Acknowledgment

This work was supported by MINECO, project number MAT2014-57704-C3. We would like to thank Tim Buscher and Prof. Armido Studer (University of Münster, Germany) for kindly providing the zeolite L crystals.

References

[1] Carrascosa M, García-Cabañes A, Jubera M, Ramiro J B and Agulló-López F 2015 *Appl. Phys. Rev.* **2** 040605.

[2] Miccio L, Marchesano V, Mugnano M, Grilli S and Ferraro P 2016 *Opt. and Lasers in Eng* **76**, 34-9.

[3] Huber S and Calzaferri G 2004 *Chem. Phys. Chem.* **5**, 239-242.

[4] Veiga-Gutiérrez M, Woerdemann M, Prasetyanto E, Denz C and De Cola L 2012 *Adv. Mater.* **24**, 5199.

[5] Muñoz-Martínez J F, Elvira I, Jubera M, García-Cabañes A, Ramiro J B, Arregui C and Carrascosa M 2015 *Opt. Mater. Express* **5**, 1137.

[6] Arregui C, Ramiro J B, Alcázar A, Méndez A, Burgos H, García-Cabañes A, and Carrascosa M 2014 *Opt. Express* **22**, 29099.

[7] Arregui C, Ramiro J B, Alcázar A, Méndez A, Muñoz-Martínez J F and Carrascosa M 2015 *J. Eur. Opt. Soc.* **10**, 15026.

IOP Conf. Series: Journal of Physics: Conf. Series **867** (2017) 012020 doi:10.1088/1742-6596/867/1/012020

Potassium Tantalate-Niobate Mixed Crystal Thin Films for Applications in Nonlinear Integrated Optics

Yuechen Jia[1,*], Markus Winkler[2], Jan Szabados[1], Ingo Breunig[1,2], Lutz Kirste[3], Volker Cimalla[3], Agnė Žukauskaitė[3], and Karsten Buse[1,2]

[1]Laboratory for Optical Systems, IMTEK, University of Freiburg, Georges-Köhler-Allee 102, 79110 Freiburg, Germany
[2]Fraunhofer Institute for Physical Measurement Techniques IPM, Heidenhofstraße 8, 79110 Freiburg, Germany
[3]Fraunhofer Institute for Applied Solid State Physics IAF, Tullastraße 72, 79108 Freiburg, Germany

*Corresponding author: *yuechen.jia@imtek.uni-freiburg.de*

Abstract. Potassium tantalate-niobate mixed crystal (KTN) thin films are promising candidates to meet the needs of integrated nonlinear optical devices for electro-optic and frequency-conversion applications. In this contribution we report on pulsed-laser-deposition growth of ferroelectric KTN films on MgO substrates. It is shown that highly-oriented KTN films are epitaxially grown as revealed by X-ray diffraction analysis. Moreover, the thermal annealing treatment can be further optimized to obtain optically smooth KTN films with RMS surface roughness as low as 1 nm.

1. Introduction
Perovskite potassium tantalate-niobate mixed crystals ($KTa_{1-x}Nb_xO_3$ with $0 \leq x \leq 1$, KTN) undergo a phase transition from a paraelectric cubic to a ferroelectric tetragonal structure with decreasing temperature [1]. By adjusting the Ta/Nb content ratio one can tune the phase-transition temperature of KTN and thus also its main properties at a given temperature. For example, temperature-dependent dielectric and electro-optic (EO) constants have been demonstrated with abrupt changes appearing around the Curie point of KTN, displaying the largest-known EO coefficients among all Kerr media [2]. Benefiting from this, this extremely promising material is showing potentials to overtake the well-known $LiNbO_3$ for EO applications such as beam scanners or modulators [2-4]. However, due to the fact that the crystals grown have compositions being different from those of the molten ingredients, high-quality and homogeneous single-crystalline KTN is difficult to produce, which is limiting the wide-spread use of this material [2]. Direct thin-film preparation, such as provided by pulsed laser deposition (PLD), appears to be an efficient way to solve the bulk-homogeneity problem. In this contribution, we report on the fabrication of smooth ferroelectric KTN thin films on MgO substrates by employing PLD and subsequent thermal annealing treatment and polishing process.

2. Experimental
The films were grown by employing a KrF excimer laser (λ = 248 nm, pulse duration of 20 ns) operating at a repetition rate of 1 Hz with a fluence of 2 J/cm^2 focused on a rotating K-enriched KTN ceramic target. The composition $x = 0.5$ is chosen because its corresponding Curie temperature ($T_c \approx$ 100 °C) ensures the crystalline structure of KTN stays in the ferroelectric phase at room temperature.

During the deposition, the substrate temperature was heated up to 700 °C, the target-substrate distance was set at 6 cm, and the oxygen atmosphere was around 0.15 mbar. Following the deposition procedure, in situ thermal post-annealing treatment at 700 °C with durations ranging from 0 to 4 hours and a subsequent chemo-mechanical polishing operation were conducted to further smoothen the KTN films. Specifically, a Logitech PM3 polishing machine and alkaline polishing suspension with 60 nm and 20 nm silicon dioxide particles (MasterMet1, MasterMet2) were used for the fine polishing process.

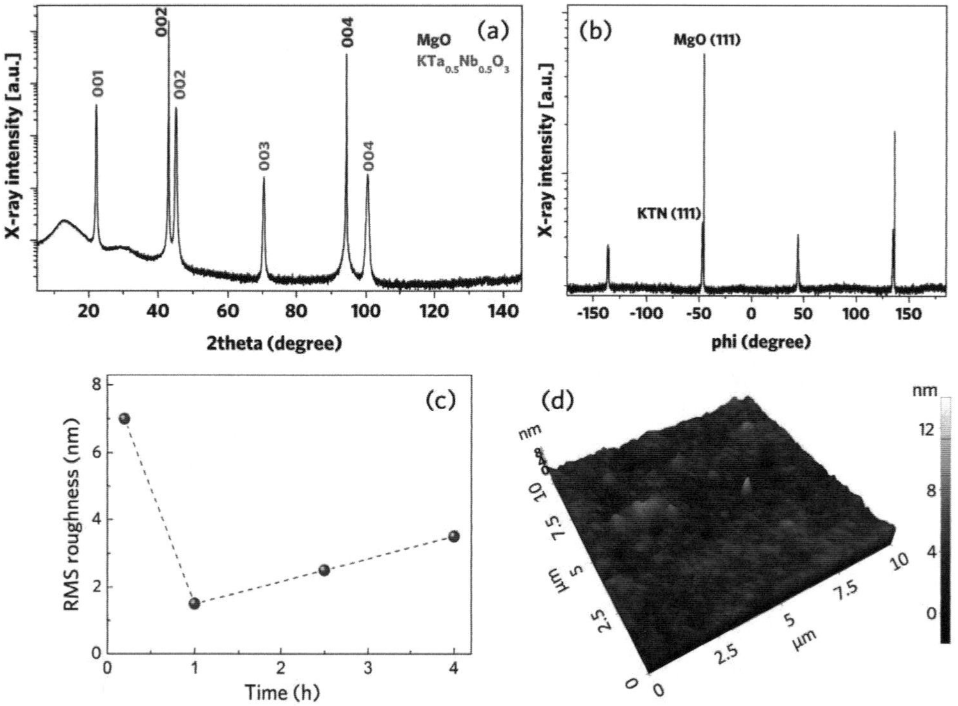

Figure 1. X-ray diffraction (a) θ-2θ and (b) φ-scan patterns of a KTN thin film deposited on (001)MgO. (c) RMS surface roughness of the polished KTN films after thermal annealing treatment with different durations. (d) AFM surface topographies of the KTN films with roughness of 1 nm.

3. Results and discussion

X-ray diffraction (XRD) analysis in θ-2θ and φ-scan modes were employed to investigate the crystalline structures of the KTN films. It was found that the as-deposited KTN films are completely (001)-oriented as shown by the θ-2θ XRD patterns in Fig. 1a, which exhibits only $00l$ peaks of KTN on (001)MgO, referring to the pseudocubic subcell. Rotationally-ordered growth of the films could be confirmed by the φ-scan XRD patterns (Fig. 1b) performed on the 111 reflection of KTN. Based on the XRD analysis, the lattice constants of the crystalline KTN thin films are determined to be $a = b =$ 3.9901 Å and c = 4.0121 Å, which are very close to those of KTN bulk crystals with tetragonal structures [5]. Moreover, energy dispersive X-ray (EDX) analysis performed on films showed that the K/(Nb+Ta) ratio is 0.95 ± 0.05 and that the Nb/Ta ratio in the films is the same as that in the starting targets, indicating good stoichiometry transfer between the target and the films. In order to obtain the surface topographies of the polished KTN films, atomic force microscopy (AFM) in contact mode was used. The change in roughness values as a function of annealing time (Fig. 1c) indicates that the duration of thermal annealing could be optimized to further reduce the RMS surface roughness to about 1 nm (Fig. 1d), which is the lowest reported value so far for PLD-grown KTN thin films. Furthermore, the waveguiding properties of the ferroelectric KTN films were studied by using the

IOP Conf. Series: Journal of Physics: Conf. Series **867** (2017) 012020 doi:10.1088/1742-6596/867/1/012020

prism coupling method (Metricon® Model 2010 prism coupler). Based on the effective refractive indices of excited TE and TM guiding modes, the refractive indices $n_o = 2.258$ and $n_e = 2.214$ of the KTN films at 632.8-nm wavelength could be determined, respectively.

The data presented in this contribution shows that PLD-grown $KTa_{0.5}Nb_{0.5}O_3$/MgO films may reach a quality satisfactory for integrated-optical applications. Our future work will focus on the fabrication and functionalization of KTN waveguide-based modulators, frequency converters and on-chip micro-resonators.

References

[1] Triebwasser S 1959 *Phys. Rev.* **4** 63
[2] Imai T, Sasaura M, Nakamura K and Fujiura K 2007 *NTT Tech. Rev.* **5** 1
[3] Nakamura K, Miyazu J, Sasaura M and Fujiura K 2006 *Appl. Phys. Lett.* **89** 131115
[4] Nakamura K, Miyazu J, Sasaki Y, Imai T, Sasaura M and F Kazuo 2008 *J. Appl. Phys.* **104** 013105
[5] Kallur V and Pandey R 1994 *Ferroelectrics* **158** 55-60

IOP Conf. Series: Journal of Physics: Conf. Series **867** (2017) 012021 doi:10.1088/1742-6596/867/1/012021

Propagation properties of the accelerating beams generated by discrete Airy-Vortex phase mask

Kun Han,[1,2] Kaiwen Ji,[1,2] Guoquan Zhang[3] and Xinyuan Qi[1,2,*]
[1]School of Physics, Northwest University, Xi' an 710069, China
[2]National Key Laboratory of Photoelectric Technology and Functional Materials (Culture Base), and Institute of Photonics and Photon-Technology, Northwest University, Xi' an 710069, China
[3]MOE Key Laboratory of Weak-Light Nonlinear Photonics, TEDA Applied Physics Institute and School of Physics, Nankai University, Tianjin 300457, China

E-mail: qixycn@nwu.edu.cn

Abstract. We present a novel type of accelerating beam generated by a discrete Airy-Vortex phase mask based on the digital holographic technology. The study shows that the main lobe and the side lobes of such beam rotate with different angular momentums and the whole beam evolves into two separated Airy-like beams in the far field. The intensity distribution of the main lobe and the side lobes in the near field can be modulated by tuning the topological gradient ΔL_1 and ΔL independently. The propagation path of the main lobe follows a parabolic trajectory. The experimental results are consistent with the numerical simulations.

1. Introduction

In recent years, the study of the accelerating beams has attracted tremendous attentions. The airy beam, a most representative accelerating beam, was first theoretically demonstrated by Berry and Balazs in 1979 and experimentally observed by Christodoulides *et al.* in 2007 [1, 2, 3]. In 2010, the projectile motion of two-dimensional truncated airy beams in a general ballistic trajectory with controllable range and height was reported by Chen *et al* [4]. Fan *et al* studied a pair of symmetrically inverted twin Airy beams in which the slope factor can regulate the spacing between the two Airy beam peaks in 2013 [5]. The accelerating quasi-Airy beam propagates along arbitrarily appointed parabolic trajectories and can be readily manipulated with tunable phase patterns, which was reported in 2016 [6]. Lately, Ren *et al* constructed and generated symmetric accelerating caustic beams which are a type of bimodal accelerating caustic beam with two quasi-constant intensity peaks, very similar to the combination of two face-to-face Airy-like beams judging by appearance [7]. All the studies made it possible for the acceleration beams to be employed in many applications, such as optical trapping [8], self-bending plasma channels [9], and electron capture and acceleration driven [10]. It is worth mentioning that a compound accelerating beam, Airy-Vortex beam, can be generated by introducing the vortex term into a Airy beam. The propagation of such compound accelerating beam in free space has been studied [11]. The results show that the vortex will affect the intensity distribution and the phase distribution. Recently, Ref. [12] reported a new type of accelerating beam generated by a discrete phase mask, in which the factors of the discrete phase mask can modulate the intensity distribution of the whole beam.

In this paper, a novel method to modulate the different parts of an accelerating beam generated by a discrete Airy-Vortex phase mask (DAVPM) based on the holographic technology is proposed. Furthermore, the propagation properties of the accelerating beam is studied numerically and experimentally.

2. Theory

To obtain the accelerating beam and modulate the main lobe, we propose a discrete Airy-Vortex phase mask(DAVPM),

$$\varphi(k_r) = \begin{cases} \exp(il_{1_{n1}}\theta), & k_R\sqrt{\frac{n_1-1}{2N}} < k_r < k_R\sqrt{\frac{n_1}{2N}} \\ \exp(ik_x^3/8 + ik_y^3/8)\exp(il_n\theta), & k_R\sqrt{\frac{n-1}{N}} < k_r < k_R\sqrt{\frac{n}{N}}, \\ 0, & \text{others}, \end{cases} \tag{1}$$

where n_1 is the layer numbers of central circle part of the DAVPM and equals to 1,2, the central part splits into two layers. The $\exp(il_{1_{n1}}\theta)$ denotes the discrete phase term of the optical vortice with different topological charges $l_{1_{n1}}$ at two layers of the central circle part, where $l_{1_{n1}} = l_{1_1} + \Delta L_1$, $l_{1_1} = 1$, ΔL_1 is the azimuthal discrete parameter representing the topological gradient and θ is the azimuthal coordinate. The $\exp(ik_x^3/3 + ik_y^3/3)$ is the cubic phase of the Fourier spectrum of Airy beam expressed in k_x-k_y space, $k_r = \sqrt{k_x^2 + k_y^2}$ and the value of k_R is set to 12. The $\exp(il_n\theta)$ denotes the discrete phase term of the rest part, which has the same meaning with the discrete phase term of the central circle part. Where N is the radial discrete parameter representing the total layer number, $l_n = 1 + (n-1)\Delta L$, the integer n is the layer number of the rest part and equals to 2 to N, the ΔL has the same meaning of ΔL_1. Consequently, ΔL_1 can modulate the central circle phase distribution of the DAVPM and ΔL can modulate the rest part.

Interfered with a plane wave, such a DAVPM can be used to fabricate a digital hologram [13, 14] with a transmission $T = \exp\{i\frac{|\varphi(k_r)|arg[\varphi(k_r)]}{max[|\varphi(k_r)|]} + iP(k_x + k_y)\}$, where P is used to separate the diffraction orders and set as 10.

Clearly, the first-order diffraction of the digital hologram, U_{F1} (in k space), contains all the information of the DAVPM after a plane wave is launched onto the hologram based on the holographic technology. According to the classical opitcs, the generation of the accelerating beam can be obtained by an inverse Fourier transform through L_3 and the propagation of the beam can be described by the Fresnel diffraction. The light field $u_1(x, y, z)$ in real space follows,

$$u_1(x, y, z) = FreT_{f+z}(FreT_f(U_{F1}) \times F), \tag{2}$$

where x and y are the transverse coordinates, z is the longitudinal coordinate, f and $F = \exp[\frac{-ik(x^2+y^2)}{2f}]$ is focal length and modulated function of the Fourier transform lens respectively, and k is the wave vector. The operator $FreT_z$ denotes the Fresnel diffraction integral from $u_0(x_0, y_0, 0)$ to $u(x, y, z)$,

$$FreT_z(u_0) = \frac{ik}{2\pi|z|}\iint u_0 \exp\{-\frac{ik}{2|z|}[(x-x_0)^2 + (y-y_0)^2]\}dxdy. \tag{3}$$

3. Experimental setup

Figure 1 shows the schematic of the experiment setup. In the experiment, a 532 nm plane wave is launched onto the spatial light modulator (SLM) through a beam splitter (BS) after expansion by lenses L_1 and L_2. The hologram which are computer calculated by the DAVPM (see the inset in Fig. 1) is exerted on the SLM. To perform the Fourier transform, the SLM is put on the front focal plane of the Fourier transform lens L_3 with a focal length of $f = 150$ mm. A spatial filter (F) on the back focal plane is used to preserve the first-order diffraction and block the other-orders. The propagation dynamics of the produced beam can be monitored by a charge-coupled device (CCD) camera.

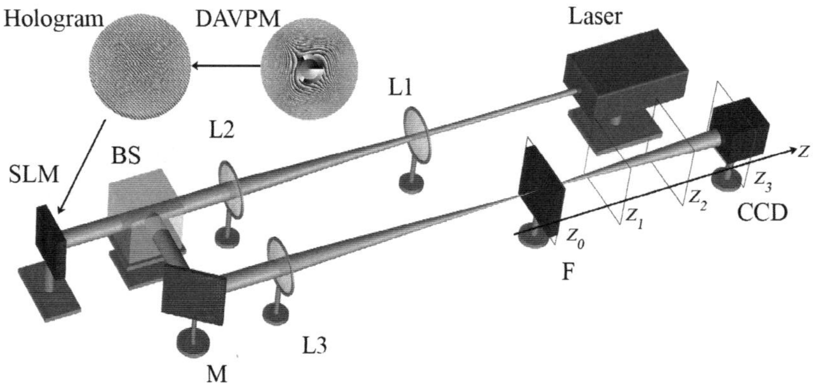

Figure 1. The schematic of the experimental setup. L, Fourier transform lens; SLM, spatial light modulator; BS, beam splitter; M, plane mirror; F, filter; CCD, charge-coupled device;

4. Experimental results and numerical simulations

4.1. Modulation to the main lobe

Figure 2 show the impact of ΔL_1 on the light field. When $\Delta L_1 = 2$, $\Delta L = 5$ and $N = 12$, the simulation (the first row) and experimental results (the second row) of the beam along the longitudinal direction z are given, respectively. At $z_0 = 0$mm, the main lobe of the produced beam is a dipole mode and the side lobes are consisted of four discontinuous arms. In the near field, the dipole mode main lobe rotates slowly just like the rotating phenomena in [15]. During the propagation along the z axis, the vortex in the main lobe vanishes gradually [see (a1)-(a4) and (b1)-(b4) in Fig. 2]. Simutaneously, the four discontinuous side lobes also rotate slowly around the main lobe. Apparently, the rotating angles $\beta_1 \neq \alpha_1$ shows that the main lobe and the side lobes rotate with different angular momentums. In the far field ($z = 81$mm), the produced beam evolves finally into two Airy-like beams [see (a6) and (b6) in Fig. 2]. The transmission trajectory of the produced beam can be described in Fig. 3. where the black solid line denotes the simulation result and the red cross symbol (same as the white cross in second row in Fig. 2) denotes the experimental results. From Fig. 2 and Fig. 3, it is obvious that the main lobe of the beam accelerates during the propagation on a parabolic path. Our simulation results are in good agreement with the experimental results.

The third and bottom rows in Fig. 2 show the simulation results and corresponding experimental results of the evolutionary patterns of the generated beam by the DAVPM with $\Delta L_1 = 3$, $\Delta L = 5$ and $N = 12$ along the longitudinal direction z, respectively. By changing ΔL_1 from 2 to 3, the main lobe changes to tripple mode which also rotates slowly during the near field while the side lobes is unchanged [see Fig. 2 (c1)-(c3) and (d1)-(d3)]. Furthermore,the produced beam also evolves into two Airy-like beams after the vortex vanished in the far field ($z = 81$mm). This is almost the same as that in the aforementioned case with $\Delta L_1 = 2$ [see Fig. 2 (c4)-(c6) and (d4)-(d6)].

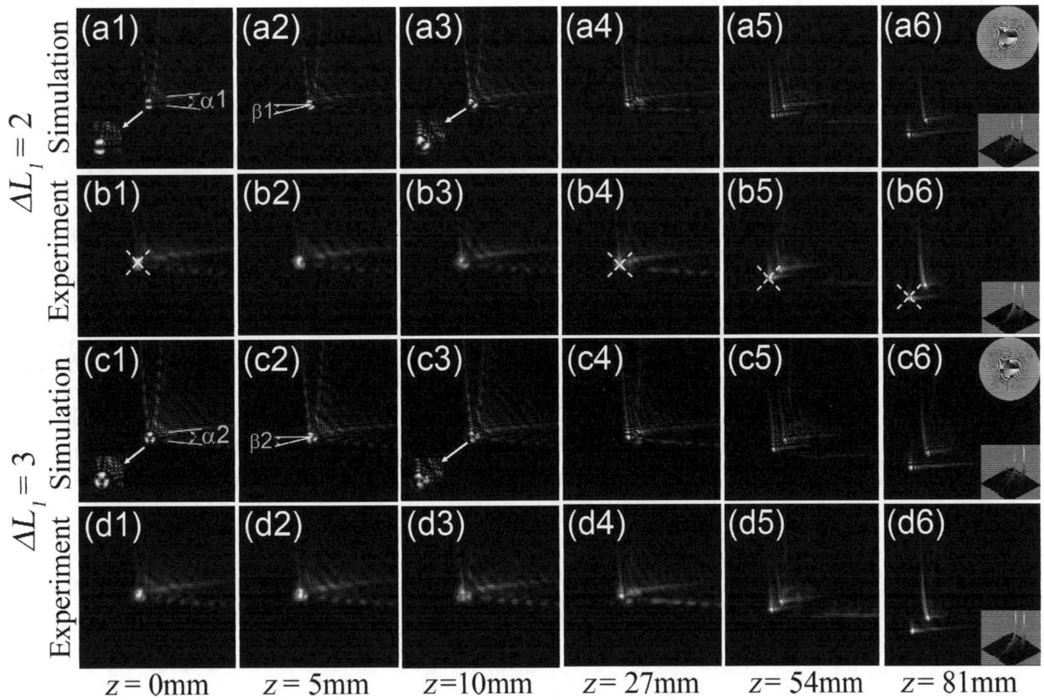

Figure 2. The evolutionary patterns of the accelerating beam along the z axis. Top row, simulations at $\Delta L_1 = 2$, $\Delta L = 5$ and $N = 12$. Top right in (a6), corresponding DAVPM. Bottom right in (a6), corresponding three dimensional field distribution at z_3. second row, corresponding experiments. Third row, simulations at $\Delta L_1 = 3$, $\Delta L = 5$ and $N = 12$. Top right in (c6), corresponding DAVPM. Bottom right in (c6), corresponding three dimensional field distribution at z_3. Bottom row, corresponding experiments.

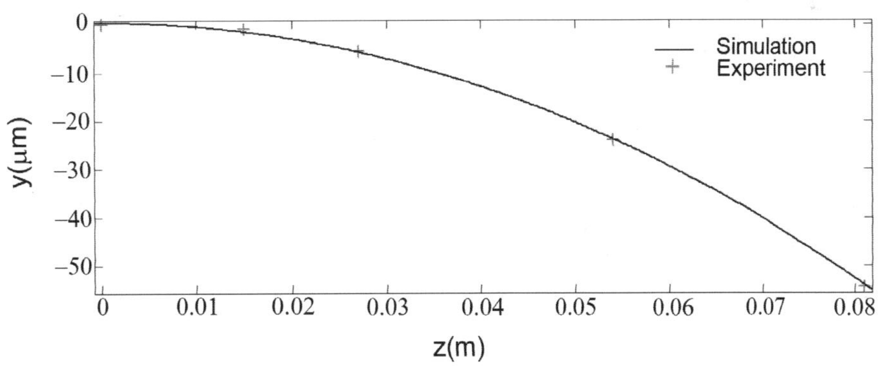

Figure 3. The trajectory curves in $y - z$ plane.

4.2. Modulations to the side lobe

Figure 4 shows the impact of ΔL on the side lobes. The parameters ΔL, N and ΔL_1 respectively equal to 2, 12, and 2. At $z = 0$mm, the opening angle between the adjacent side lobes $\alpha 3$ decrease clearly compared with $\alpha 1$ and $\alpha 2$. In the near field, the rotation angle $\beta_3 \approx \beta_1$, denoting that the rotation angle of the dipole mode main lobe is independent on ΔL [see Fig. 4 (a1)-(a3) and Fig. 2 (a1)-(a3)]. Furthermore, after the vortex of the main lobe vanished in the far field ($z = 81$mm), the produced beam finally evolves into two Airy-like beams, too. Note that the light intensity of the inner Airy-like beam is pretty stronger than that of the outer one due to the impact of the smaller ΔL [see Fig. 4 (a4)-(a6) and (b4)-(b6)].

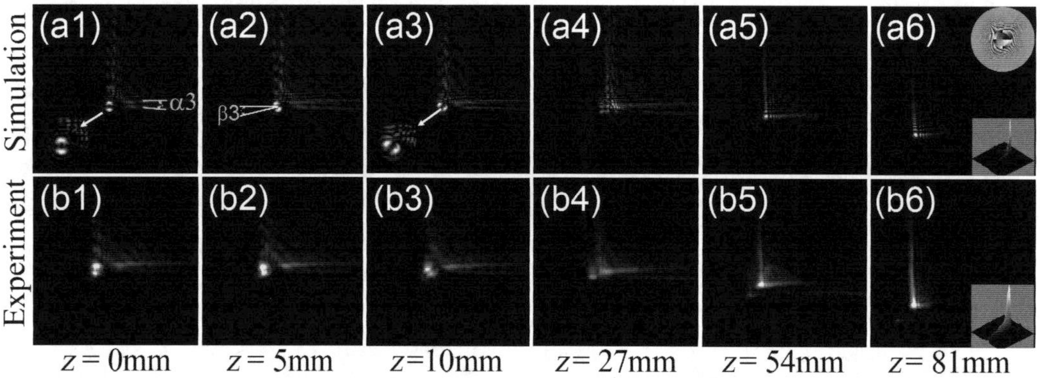

Figure 4. The evolutionary patterns of the accelerating beam with $\Delta L_1 = 2$, $\Delta L = 2$ and $N = 12$ along the z axis. Top row, simulations. Top right in (a6), corresponding DAVPM. Bottom right in (a6), corresponding three dimensional field distribution at z_3. Bottom row, corresponding experiments.

5. Conclusion

In conclusion, we have proposed a novel kind of accelerating beam which can be generated by a tunable DAVPM with holographic technology. In the near field, the main lobe and the side lobes rotate slowly with different angular momentums. Such accelerating beam finally split into two separated Airy-like beams in the far field ($z = 81$mm). The whole propagation of the main lobe follows along a parabolic trajectory. Particularly, the intensity distribution of the main lobe in the near field can be controlled by tuning the factor ΔL_1 of the DAVPM, and the opening angle between side lobes can be modulated by the factor ΔL. Furthermore, after the vortex of the main lobe vanished in the far field ($z = 81$mm), the factor ΔL mainly modulates the intensity distribution of two airy-like beams. Our experimental observations are in accord with the numerical simulations. The intriguing characteristics of the accelerating beam may open up new avenues for optical field manipulation.

Acknowledgment

The work was supported by the National Natural Science Foundation of China (11104221) and Open fund of MOE Key Laboratory of Weak-Light Nonlinear Photonics(OS17-3).

References

[1] M. V. Berry and N. L. Balazs 1979 *Am. J. Phys* **47** 264-7
[2] G. A. Siviloglou, J. Broky, A. Dogariu, and D. N. Christodoulides 2007 *Phys. Rev. Lett* **99** 213901
[3] G. A. Siviloglou and D. N. Christodoulides 2007 *Opt. Lett* **32** 979-81
[4] Y. Hu, P. Zhang, C. Lou, S. Huang, J. Xu, and Z. Chen 2010 *Opt. Lett* **35** 2260-2
[5] Y. T. Fan, J. S. Wei, J. Y. Ma, and Y. Q. Wu 2013 *Opt. Lett* **38** 1286-8
[6] Y. X. Qian, and S. Zhang 2016 *Opt. Express* **24** 9489-500
[7] Z. J. Ren, and H. Z. Jin, B. J. Peng and Y. L. Shi 2016 *Appl. Opt* **55** 7694-9
[8] J. Baumgartl, M. Mazilu, and K. Dholakia 2008 *Nat. Photon* **2** 675-8
[9] F. Bleckmann, A. Minovich, J. Frohnhaus, D. N. Neshev, and S. Linden 2013 *Opt. Lett* **38** 1443-5
[10] J. X. Li, W. P. Zhang, and J. G. Tian 2010 *Opt. Lett* **35** 3258-60
[11] M. Mazilu, J. Baumgartl, T. Cizmar, and K. Dholakia 2009 *Proc. SPIE* **7430** 74300C-1

[12] Kun Han, Shasha Li, Kaiwen Ji, Zengrun Wen, Meng Luo, Pingan Gao, Guoquan Zhang and Xinyuan Qi "Observation of the accelerating beams generated by the discrete Airy-Vortex phase mask." Submitted to *Optics Express*
[13] J. A. Davis, D. M. Cottrell, J. Cmpos, M. J. Yzuel, and I. Moreno 1999 *Appl. Opt* **38** 5004-13
[14] Y. V. Izdebskaya, T. H. Lu, D. N. Neshev, and A. S. Desyatnikov 2014 *Appl. Opt* **53** B248-53
[15] M. Baranek, and Z. Bouchal 2014 *Proc. of SPIE* **9441** 94410N-1

IOP Conf. Series: Journal of Physics: Conf. Series **867** (2017) 012022 doi:10.1088/1742-6596/867/1/012022

Controlling light in Airy and higher-order caustic photonic structures

Alessandro Zannotti, Falko Diebel, Matthias Rüschenbaum, Cornelia Denz

Institute of Applied Physics and Center for Nonlinear Science (CeNoS), University of Muenster, 48149 Muenster, Germany

E-mail: a.zannotti@uni-muenster.de

Abstract. Caustics form geometrically stable structures in light and are hierarchically categorized by the catastrophe theory. We embed higher-order cusp and swallowtail catastrophes in paraxial beams and investigate their dynamics. Utilizing high-intensity caustics that propagate on curved trajectories, we realize photonic caustic lattices in photosensitive media, and demonstrate waveguiding with a rich diversity of light guiding paths.

1. Introduction

Light reflected by curved surfaces or refracted by refractive index modulations shaped according to nonlinear potential functions forms caustics as geometrically stable structures whose characteristics are predicted by an analysis of these potentials in terms of bifurcation and catastrophe theory [1, 2].

The description of catastrophes within a general bifurcation theory is valid for all areas of nonlinear dynamic systems and describes their abrupt changes due to perturbations when changing external control parameters. In optics, the manifestation of catastrophes as caustics is defined by the envelope of rays, i.e. a discontinuous change of the number of rays that cross in each point of space. Optical catastrophes are classified by the number of state and control parameters, and can be found in nature or created artificially. For a single control parameter, the only structurally stable catastrophe is the fold catastrophe, which exhibits the Airy function shown as a diffraction pattern [3] in Fig. 1 (a1) and as ray optics picture in (a2). Whereas in the fold catastrophe two trajectories coalesce in a singular point, higher-order catastrophes, as the cusp catastrophe, emerge when three merging trajectories lead to the caustic Pearcey beam [4], shown in Fig. 1 as transverse intensity distribution (b1) and as corresponding ray optics picture (b2).

2. Analyzing the propagation of higher-order optical catastrophes

In recent years, caustic beams have experienced a renaissance in optics fostered by spatial light modulator techniques. With these, both the fold catastrophe-based Airy beam and the cusp catastrophe-based Pearcey beam have been realized as paraxial beams. Striking properties of these beams include acceleration in the transverse direction, auto-focusing, and form-invariant propagation [4]. The experimentally obtained propagation of the Pearcey beam is depicted in Fig. 1 (d).

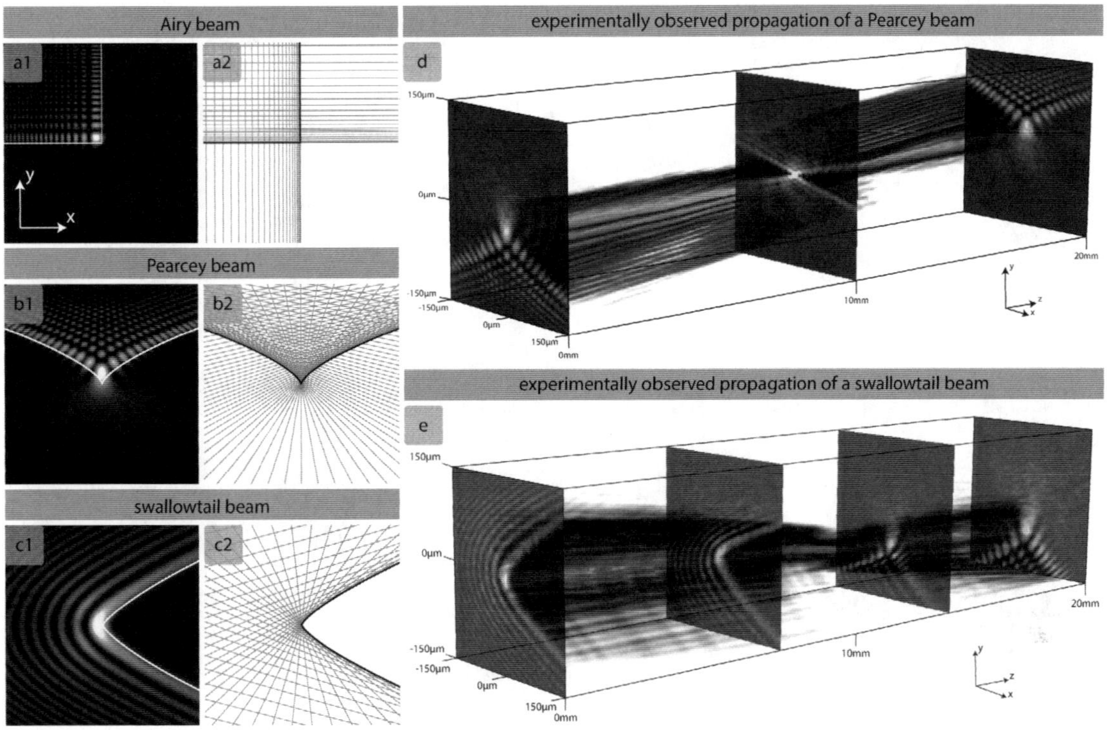

Figure 1. Catastrophes form caustics in light. (a) fold, (b) cusp, (c) swallowtail catastrophe as (1) transverse intensity patterns and (2) corresponding ray pictures. Experimentally observed propagation of a (d) Pearcey beam and (e) swallowtail beam.

We demonstrate our investigations on embedding higher-order catastrophes as e. g. the swallowtail, butterfly, and hyperbolic umbilic catastrophes in spatially structured light fields [5]. Fig. 1 (c) shows a transverse swallowtail intensity pattern (c1) and the corresponding rays (c2). We examine experimentally and theoretically the propagation of these caustic beams [6]. Exemplary, Fig. 1 (e) shows the decay of the unstable swallowtail caustic to a cusp caustic that is stable in the two-dimensional transverse plane.

3. Nonlinear photonic caustic lattices

Mapping catastrophes to light to tailor these spatial caustics experienced explosive growth during the past decade and became a vital branch in modern photonics. However, linking nonlinear photonics, especially artificial nonlinear refractive index changes, with these artificial caustic light fields is still a challenge. Combining both topics opens path breaking new aspects in information storage, guiding, and processing, and enables light localization.

In this spirit, we utilize these higher-order cusp and swallowtail catastrophes in paraxial light to transform them into photonic refractive index caustic lattices in nonlinear media [7] as shown in Fig. 4. We demonstrate waveguiding with a rich diversity of light guiding paths along 2D quasi cubic Fig. 4 (a) and 3D curved caustic structures Fig. 4 (b), and present their functionality as e.g. optical splitters Fig. 4 (c). Finally, taking advantage of the strong auto-focusing of Pearcey beams, we demonstrate the formation of a Pearcey solitary wave Fig. 4 (d) in focusing nonlinearity [7].

IOP Conf. Series: Journal of Physics: Conf. Series **867** (2017) 012022 doi:10.1088/1742-6596/867/1/012022

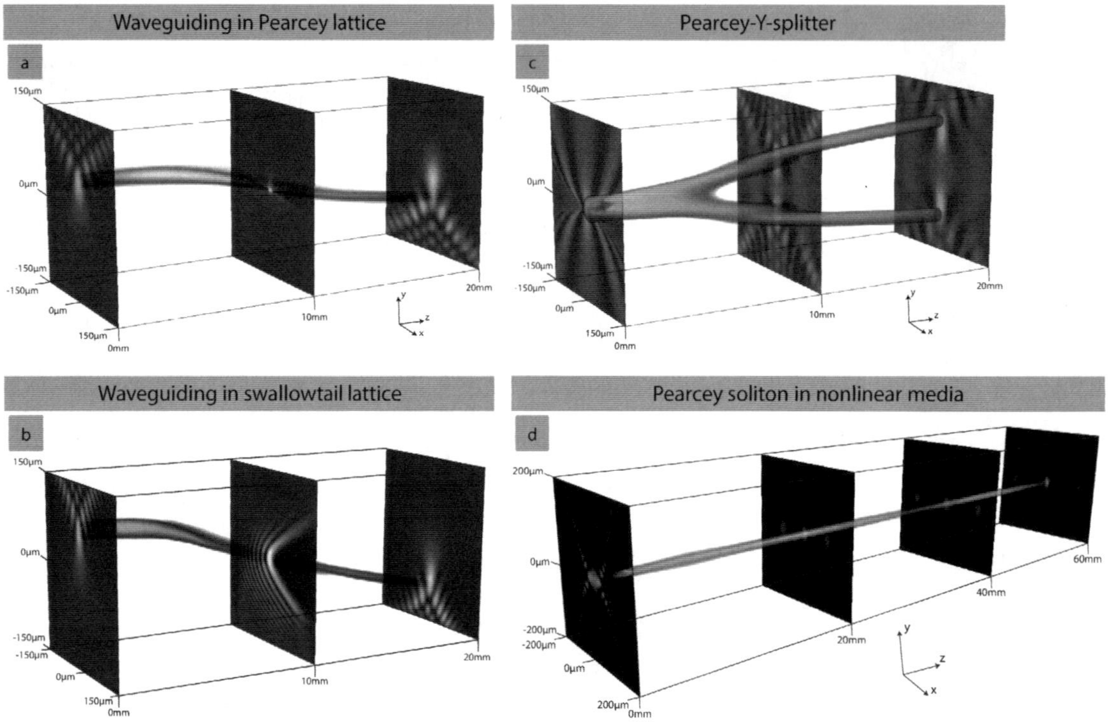

Figure 2. Waveguiding in caustic lattices and soliton formation. (a) A Pearcey lattice (gray) features light guiding (red) on 2D quasi cubic paths. (b) A Swallowtail lattice provides 3D curved waveduides. (c) Combining caustic beams enables the fabrication of a functional Pearcey-Y-splitter. (d) Nonlinear formation of a Pearcey soliton (gray & red) by exploiting the strong auto-focusing of two Pearcey beams.

4. Conclusion

To summarize, our work represents the first realization of higher-order photonic caustic lattices, and paves the way to apply this approach to a manifold of applications ranging from linear waveguiding to nonlinear soliton formation. Within the class of caustic beams diverse light structures exist that each exhibit unique propagation properties which can be combined to more complex light fields. Thus, our work provides a toolbox for versatile applications in micro-manipulation, material and signal processing, and microscopy.

References

[1] Arnold V I and Wassermann G S 2003 *Catastrophe Theory* 3rd ed (Berlin Heidelberg New York: Springer Verlag)
[2] Berry M V and Upstill C 1980 *Prog. in Optics* **18** 257–346
[3] Siviloglou G A, Broky J, Dogariu A and Christodoulides D N 2007 *Phys. Rev. Lett.* **99** 213901
[4] Ring J D, Lindberg J, Mourka A, Mazilu M, Dholakia K and Dennis M R 2012 *Opt. Express* **20** 18955
[5] Zannotti A, Diebel F, Boguslawski M and Denz C 2017 *New J. Phys.* **19** 053004
[6] Zannotti A, Diebel F and Denz C 2017 (*Preprint* `1705.00647`)
[7] Zannotti A, Rüschenbaum M and Denz C 2017 *J. Opt.* (*Preprint* `doi:10.1088/2040-8986/aa77d9`)

IOP Conf. Series: Journal of Physics: Conf. Series **867** (2017) 012023 doi:10.1088/1742-6596/867/1/012023

Formation of Photonic Structures in Photorefractive Lithium Niobate by 1D and 2D Bessel-like Optical Fields

A Inyushov, P Safronova, I Trushnikov, A Sarkyt, and V Shandarov

Tomsk State University of Control Systems and Radioelectronics, Lenin Prospect 40, Tomsk, 634050, Russia

E-mail: shandarovvm@svch.rk.tusur.ru

Abstract. Both, one-dimensional (1D) and two-dimensional (2D) Bessel-like beams with different topology of 2D beam cross-sections are formed from Gaussian laser beams using the amplitude masks and Fresnel biprisms. These almost diffraction-free light fields with wavelengths of 532 and 633 nm can change the refractive indices of photorefractive lithium niobate samples and form within them the nonlinear photonic diffraction structures. The characteristics of photonic structures induced in this way are studied by diffraction of monochromatic light with wavelengths of 633 and 532 nm.

1. Introduction

In the most cases any laser source generates Gaussian-shaped light beam which may be tightly focused with beam waist region in longitudinal direction depending on the light wavelength and minimal transverse waist size. However, some applications require not conventional shapes of light beams which demonstrate some properties distinct of Gaussian laser beams. These are diffraction-free beams including Bessel-like ones [1], Airy beams [2] and some other non-diffracting shape-preserving light beams [3]. The Bessel-like beams are close to theoretical diffraction-free fields which are not limited in the transverse directions. The real Bessel beams cannot exist because of the infinite optical power they should carry. However, there are some configurations that may form Bessel-like beams in the bounded space area.

The usual ways to form two-dimensional almost diffraction-free light fields exploit so called axicon lenses, annular apertures or optical fiber elements [1, 5]. However, in some cases not only two-dimensional light fields are required but also one-dimensional ones. Strictly speaking they are quasi-one-dimensional Bessel-like beams because in the second transverse dimension these fields display the Gaussian-like profiles if they are formed by laser beams.

The main aim of this study is formation of one-dimensional and two-dimensional Bessel-like fields with different shapes of transverse light patterns using diffraction grating-like amplitude transparencies and Fresnel biprisms. The obtained longitudinally homogeneous light patterns generate photonic structures, e.g. waveguide or diffraction systems in photorefractive lithium niobate ($LiNbO_3$) samples [6, 7]. We use for this purpose of amplitude masks including their couples rotated with respect to each other at some angles. Every mask contains the metal screen with two rectangular slits in it.

2. Experimental conditions and experimental results

The solid-state YAG:Nd^{3+} laser with light wavelength of λ=532 nm and He-Ne laser (λ=633 nm) are used as CW light sources in experiments. The near to parallel laser beam illuminates the biprism or amplitude mask. The amplitude mask is located in the focal plane of a lens (cylindrical or spherical, depending on one-dimensional or two-dimensional light field is formed). The longitudinally uniform interference pattern appears immediately after biprism due to interference of Gaussian beam halves passed through the prism (Figure 1 a). In the scheme with amplitude mask two light beams produced by the slits in a screen interfere in an area after lens (Figure 1 b). To form two-dimensional Bessel-like field, we use two amplitude masks with required angle between directions of their slits. Formed interference patterns generate photonic lattices or phase diffraction gratings in the photosensitive material. We use for this purpose the photorefractive samples $LiNbO_3$:Fe and $LiNbO_3$:Cu. To test properties of the phase photonic structure generated within the sample, we use laser radiation with λ=633 nm and λ=532 nm. The near field and far field diffraction patterns are studied with a CCD camera at this stage using as the unfocused as the tightly focused laser beams.

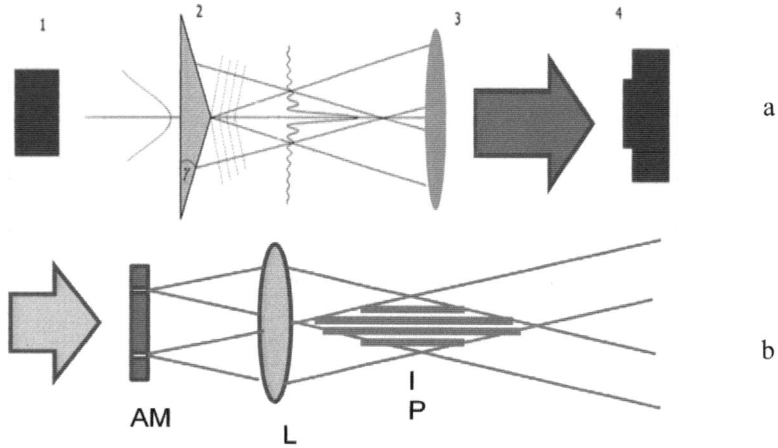

Figure 1. Experimental setups for formation of Bessel-like beams with Fresnel biprism (a), and with amplitude masks (b). 1 – laser source, 2 – biprism, 3 – imagine lens, 4 – beam analiser, AM – amplitude mask, L – lens.

2.1. Bessel-like light beam formation with amplitude masks

As the particular results, images in Figure 2 show 1D (λ=633 nm) and 2D light patterns (λ=532 nm) near the back focal plane of a cylindrical lens with focal length 80 mm (a), and spherical lenses (b - d) with focal length of 190 mm at different dimensions of slits and different slit orientation with respect to each other. To form 1D beam, we use the metal mask with two parallel rectangular slits. The slit cross-section measures 0.1×1.5 mm^2 with a distance between slit centers of 0.6 mm. Similar masks are used to form 2D Bessel-like beams with slit width 0.2 mm and their length 2 mm. The distances between their centers range from 0.5 to 1 mm.

The interference nature of formed light fields in the experiments provides their much better longitudinal homogeneity compared with 2D Gaussian beam. It is illustrated by images in Figure 3 with dependences of intensity cross-sections of Gaussian beam and 2D interference field on position of CCD camera along the light propagation direction measured in centimeters. Figure 4 demonstrates corresponding dependences of Gaussian beam diameter at half maximum intensity level and the width of interference field central maximum for the case when they are comparable in dimensions. It is clearly seen that variation of transverse dimensions of this Bessel-like beam is much less when compared with that for the usual Gaussian beam.

IOP Conf. Series: Journal of Physics: Conf. Series **867** (2017) 012023 doi:10.1088/1742-6596/867/1/012023

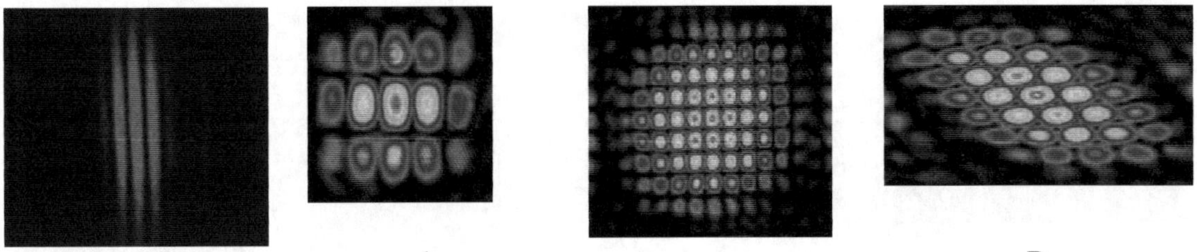

a b c D

Figure 2. Light cross-sections of 1D and 2D Bessel-like fields for horizontal widths of central maxima: a – 20 μm; b – 230 μm (90° angle between slit directions); c – 60 μm (the same angle); c – 90 μm (45° angle between slit directions).

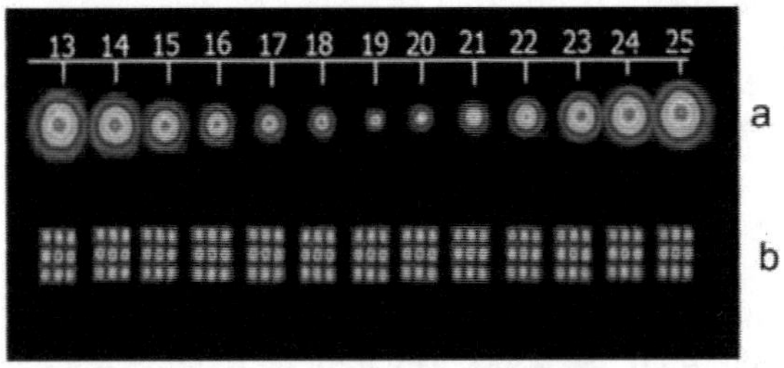

Figure 3. Intensity cross-sections of Gaussian beam (a) and 2D interference field (b) at different positions along light propagation direction.

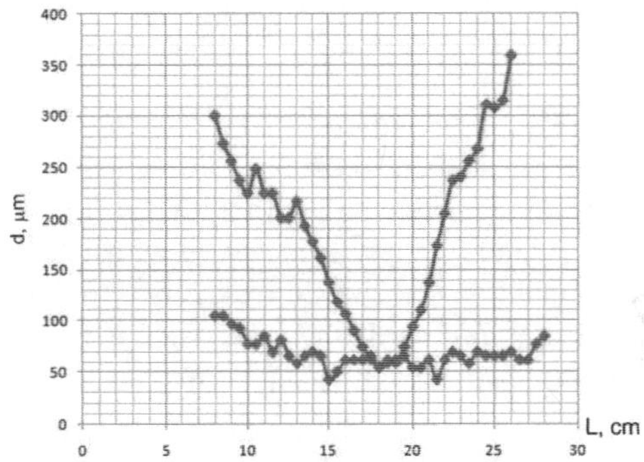

Figure 4. Gaussian beam diameter (red) and the central maximum diameter of interference pattern (blue) along light propagation direction.

2.2. Bessel-like light beam formation with Fresnel biprisms

Two-dimensional Bessel-like beams are formed in usual cases with laser beam passed through conical lenses, so-called «axicons» [1]. To form quasi-one-dimensional Bessel-like beams it is possible to exploit Fresnel biprisms [1] which can be considered as 1D analog of axicon lens. A laser beam with Gaussian profile is separated by a prism into two parts which interfere with each other in a space after the prism. The transverse profile of interference pattern in this case is close to that for Bessel beam. The spatial "period" of interference image Λ is determined in this case by expression

$$\Lambda = \frac{\lambda}{2\sin((n-1)\gamma)} \qquad (1),$$

where λ is the light wavelength, n is the refractive index of prism material, and γ is the angle of the prism (Figure 1 a).

We use biprisms with these angles 4° and 12° in experiments. Images in Figure 5 illustrate light pictures in cross/sections of laser beam (λ=532 nm) passing through the prism with 4° angle in different positions. Picture a shows beam before prism, pictures b and c demonstrate cross-sections of light beams passed through the prism after the interference region, and picture d shows interference light pattern immediately after prism. It should be noted that light field is strongly scattered at diffraction on the apex line of the prism [8] and the scattering band is seen between light beam halves (Figure 5 b). To suppress this scattering, the prism apex is screened by black paper band and scattering disappears. It increases the contrast of interference pattern (Figure 5 d).

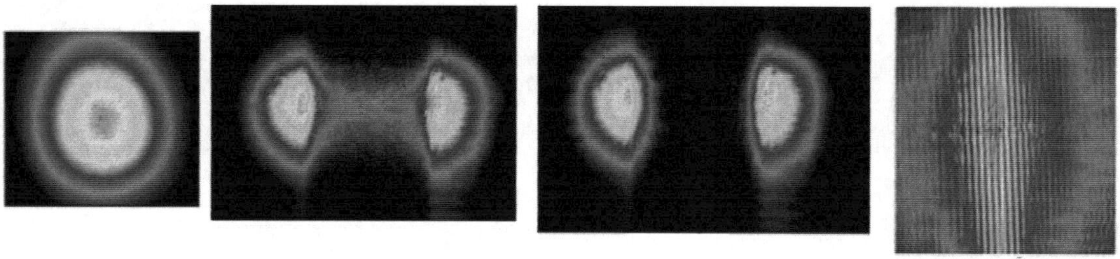

a b c d

Figure 5. Cross-sections of laser beam (λ=532 nm) passing through biprism. Initial laser beam (a), light picture outside the interference area with scattering at the prism apex (b), the same image with apex screened (c), and interference pattern immediately after prism (d).

2.3. Formation of photonic structures by Bessel-like beams in photorefractive lithium niobate

Bessel-like 1D and 2D beams are used to generate phase diffraction structures in photorefractive samples of copper-doped plate $LiNbO_3$:Cu (0.05 wt%). The dimensions of this plate are 10×2×15 mm^3 along X, Y, and Z axes. It is doped with Cu while crystal growth. The light polarization at phase structure induction corresponds to the crystal extraordinary or ordinary waves. At the structure readout we use the extraordinarily polarized light waves of YAG or He-Ne lasers. The light pattern in Figure 6 a illustrates near-field diffraction of light (λ=532 nm) on the 1D few-element phase grating with spatial period 180 μm induced within the $LiNbO_3$:Cu plate. The optical power and exposure time at this grating creation are 1 mW and 3 minutes. Image in figure 6b shows cross-section of 2D interference pattern obtained with two amplitude masks 90° rotated with respect to each other in the same crystal plate. And image in Figure 6c demonstrates near field diffraction pattern at the exit facet of that plate obtained with laser beam (λ=532 nm) diffraction at the phase grating induced in $LiNbO_3$:Cu plate. It should be noted that exposure time to induce this 2D structure makes up 30 minutes at the same light power as it was used at 1D grating formation.

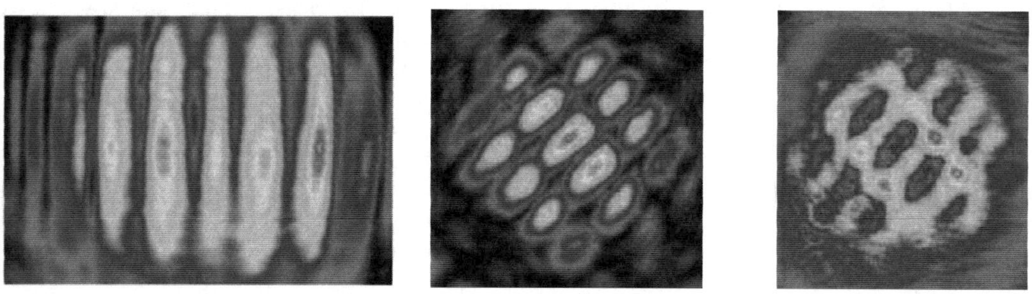

a b c

Figure 6. Cross-sections of laser beam (λ=532 nm) passing through biprism. Initial laser beam (a), light picture outside the interference area with scattering at the prism apex (b), the same image with apex screened (c), and interference pattern immediately after prism (d).

Images in Figure 7 demonstrate the change of 1D diffraction grating profile in time during its generation in photorefractive plate $LiNbO_3$:Cu. They are obtained with amplitude mask scheme. The spatial period of phase grating in this case is 100 μm. At the initial stage of the grating formation (less than 10 seconds) the near field diffraction image and its intensity profile are practically the same as in the light field. However, at the exposure time increase (t=80 s) the arising of maximal intensity within side lobes of near field diffraction image as well as in intensity profile of that image are observed.

Indeed, the photorefractive optical nonlinearity is saturable and that results in possible creation of phase diffraction and waveguide elements with required profiles in photorefractive lithium niobate.

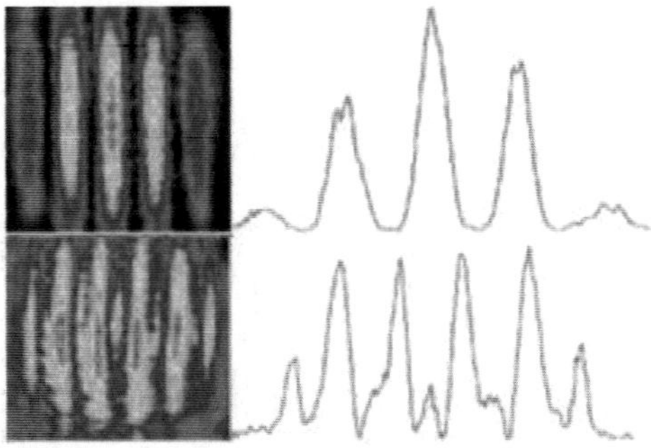

Figure 7. Near field diffraction images at the exit surface of $LiNbO_3$:Cu plate (left) and their intensity profiles (right) for exposure times less than 10 s (upper line) and 80 s (lower line).

3. Conclusion

In conclusion, our experimental results confirm the possible creation of few-element phase diffraction structures and more complicated photonic waveguide circuits with required profiles within photosensitive materials like photorefractive crystals by light fields with Bessel-like shapes. Variation of parameters of optical schemes gives the additional degree of freedom to create Bessel-like light beams with needed characteristics.

Acknowledgments

This study was carried out with the financial support of Ministry of Education and Science of Russia (within the task N 3.1110.2017/PCh of the project part).

References

[1] Duocastella M and Arnold C B 2012 *Laser Photonics Rev.* **6** 607nother reference
[2] Siviloglou G A and Christodoulides D N 2007 *Optics Letters* **32** 979
[3] Bandres M A and Rodrıguez-Lara B M 2013 *New Journal of Physics* **15** 013054
[4] Henderson D M and Abrams R L 1970 *Optics Communications* **5** 223
[5] Zhu X, Schülzgen A, Li L, and Peyghambarian N 2009 *Applied Physics Letters* **94** 01102
[6] Kip D 1998 *Appl. Phys. B* **67** 131
[7] Petrov M P, Stepanov S I and Khomenko A V 1991 *Photorefractive Crystals in Coherent Optical Systems* (Berlin: Springer-Verlag)
[8] Lei M, Yao B 2004 *Optics Communications* **239** 367

Quantum-like adiabatic light transfer in photo-induced waveguides with longitudinally varying detuning

Hassan Oukraou[1,2], Laura Vittadello[1,3], Virginie Coda[1,2], Charles Ciret[1,4], Massimo Alonzo[1,5], Andon A. Rangelov[6], Nikolay V. Vitanov[6], and Germano Montemezzani[1,2]

[1] Université de Lorraine, Laboratory LMOPS, Metz, France
[2] CentraleSupélec, Laboratory LMOPS, Metz, France
[3] Dipartimento di Fisica e Astronomia, Università di Padova, Padova, Italy
[4] OPERA-Photonique, Université Libre de Bruxelles (ULB), Bruxelles, Belgium
[5] Ultrafast Photonics Laboratory, Sapienza Università di Roma and CNISM, Roma, Italy
[6] Department of Physics, Sofia University, Sofia, Bulgaria

E-mail: hassan.oukraou@centralesupelec.fr

Abstract. Besides longitudinally varying coupling constants, the longitudinal variation of the propagation constants leads to an additional parameter for the control of adiabatic light transfer in coupled waveguide systems. Examples are given using waveguides structures recorded with the help of the photorefractive effect and mimicking the quantum processes of Rapid Adiabatic Passage (RAP) and two-state STImulated Raman Adiabatic Passage (two-state STIRAP).

1. Introduction

Photorefractive light-induced waveguides are a useful platform for testing of novel functionalities in integrated optics inspired by their analogies with quantum physical processes as well as for the "classical" verification of involved quantum analogies. The latter stem from the fact that waves passing from a waveguide to another via evanescent coupling represent a classical analogous to quantum population dynamics processes between levels being resonantly or non resonantly coupled by an appropriate exciting field [1, 2]. Both processes are described by a Schrödinger-type equation. In this context we use an approach based on lateral illumination of a properly shaped control-light in an electrically biased $Sr_{0.61}Ba_{0.39}Nb_2O_6$ (SBN) crystal. Our earlier studies concerned structures composed of waveguides with an identical propagation constant β (equivalent to a resonant transfer) and with adiabatically evolving coupling constants between them. For instance, systems mimicking the "STImulated Raman Adiabatic Passage" (STIRAP) in a record multistate system composed of 11 waveguides [3], or systems implementing novel broadband multiple beam splitters were demonstrated experimentally [4]. Here we discuss the role of an additional control parameter, the detuning of the propagation constants (proportional to the effective refractive indices) of the involved waveguides. Introducing an adiabatic evolution of this parameter leads to rich new functionalities. Examples are waveguide structures related to the quantum processes of "Rapid Adiabatic Passage" (RAP) and "Two-state STIRAP" leading respectively to broadband directional couplers and broadband 50:50 beam splitters [5].

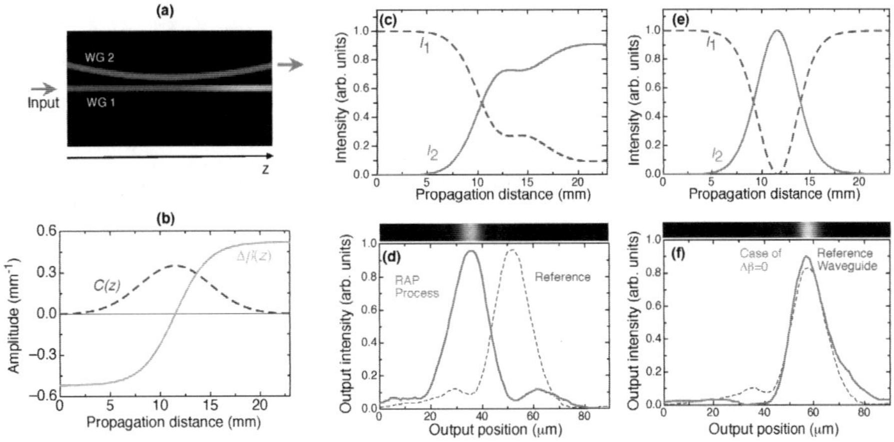

Figure 1. (a) Two-waveguide configuration for RAP-like light transfer; (b) Theoretical spatial evolution of coupling constant $C(z)$ and detuning $\Delta\beta(z)$; (c) Expected intensity evolution in the two waveguides; (d) Experimental output profile, the reference is in absence of WG 2; (e) and (f), same as (c) and (d) but for counterexample with $\Delta\beta(z) = 0$.

2. Example: Light transfer by Rapid Adiabatic Passage

We limit ourselves to a brief description of adiabatic light transfer in a structure that mimics the RAP process in a two-level quantum system [6], where the transition detuning crosses zero while the coupling pulse is applied. RAP leads to a robust and rapid way to adiabatically invert the two-state system. Figure 1(a) illustrates the optical analogous in a two-waveguides system for which the propagation constant β of one of the waveguides varies smoothly and its value crosses the one of the other waveguide at half propagation distance. At the same time the coupling constant C responsible for the evanescent light coupling has a maximum at half propagation (shortest distance between the waveguides) and diminishes towards the two ends (Fig. 1(b)). By integration of the appropriate coupled equations [5], the expected evolution of the light intensity in the two waveguides can be calculated (Fig. 1(c)), leading essentially to a transfer to the second waveguide for a broad wavelength range. Our photorefractive platform for the experimental implementation is described in [3, 4], the transfer to the second waveguide for the propagating wavelength of 633 nm is seen in Fig. 1(d), that shows the light profile at the output of our 23 mm long SBN crystal hosting the waveguides. The reference profile corresponds to the case where only the input waveguide is present. A similar transfer is observed also for propagation at the wavelength of 850 nm in the same waveguide structure, which prove the robustness of the approach. Finally, Fig. 1(e) and Fig. 1(f) show the simulations and results for the counterexample where there is no detuning of the propagation constants ($\Delta\beta = 0$). In this case no RAP-like process occurs and the results strongly depend on the probe wavelength. This proves the essential role played by the detuning $\Delta\beta$ on the RAP-like broadband light transfer.

3. References

[1] Paspalakis E 2006 *Optics Commun.* **258** 30
[2] Longhi S, Della Valle G, Ornigotti M and Laporta P 2007 *Phys. Rev.* B **76** 201101
[3] Ciret C, Coda V, Rangelov A A, Neshev D N and Montemezzani G 2013 *Phys. Rev.* A **87** 013806
[4] Ciret C, Coda V, Rangelov A A, Neshev D N and Montemezzani G 2012 *Opt. Lett.* **37** 3789
[5] Oukraou H, Vittadello L, Coda V, Ciret C, Alonzo M, Rangelov A A, Vitanov N V and Montemezzani G 2017 *Phys. Rev.* A **95** 023811
[6] Vitanov N V, Halfmann T, Shore B W and Bergmann K 2001 *Annu. Rev. Phys. Chem.* **52** 763

IOP Conf. Series: Journal of Physics: Conf. Series **867** (2017) 012025 doi:10.1088/1742-6596/867/1/012025

The analyzation of 2D complicated regular polygon photonic lattice

Jing Lv, Yuanmei Gao *

College of Physics and Electronics, Shandong Normal University, Jinan 250014, PR China

E-mail address: gaoyuanmei@sdnu.edu.cn

Abstract. We have numerically simulated the light intensity distribution, phase distribution, far-field diffraction of the two dimensional (2D) regular octagon and regular dodecagon lattices in detail. In addition, using the plane wave expansion (PWE) method, we numerically calculate the energy band of the two lattices. Both of the photonic lattices have the band gap. And the regular octagon lattice possesses the wide complete band gap while the regular dodecagon lattice has the incomplete gap. Moreover, we simulated the preliminary transmission image of photonic lattices. It may inspire the academic research both in light control and soliton.

1. Introduction

In these decades, photonic lattices have attracted substantial research interests. More and more research achievements are reported about new lattice fabrication, light control [1], localized transmission and energy band gap. Especially the band gaps, which opens a new door for studying the optical phenomenon of period photonic lattices. It is easier to find the localized transmission of the period lattice whose frequency is close to a bang gap.

As we know, a great number of studies report on the band structure of two dimensional (2D) simple period photonic lattices such as hexagon, honeycomb [2] and kagome [3] lattice by now [4]. Hexagon lattice has a range of application in nonlinear optic fiber and optical soliton transmission due to their wide band gaps [5]. The lattice solitons transmission, localization and interband transitions have been studied in honeycomb lattices [6]. Moreover, the kagome lattice has been generated by the technique of optical induction [7,8].

In this paper, we mainly discussed a new kind of structure lattices, the 2D complicated period regular polygon photonic lattice. We analysed the light field of photonic lattice by simulating the light intensity distribution, phase distribution as well as far-field diffraction pattern. For further studies, we calculated the energy band gap of it via the plane wave expansion (PWE) method in TE mode and TM mode. The periodic dielectric equation of photonic crystal and Maxwell equations are the major components. Also we need make a Fourier series expansion for the periodic function. And, the band gap has been found in the lattices whose shapes are regular octagon and regular dodecagon via calculation. In order to analysis the band structure briefly, we have simulated the image of Gauss source passing through the photonic crystal by means of the Finite-Difference Time-Domain(FDTD).

2. Analyzation of the light field

We analysed two types of 2D period photonic lattices, one is regular octagon lattice, the other is regular dodecagon lattice. The lattice is a superposition of simple lattices with a certain angle and different phases [9]. And we simulated the light intensity distribution, phase distribution and far-field diffraction of them.

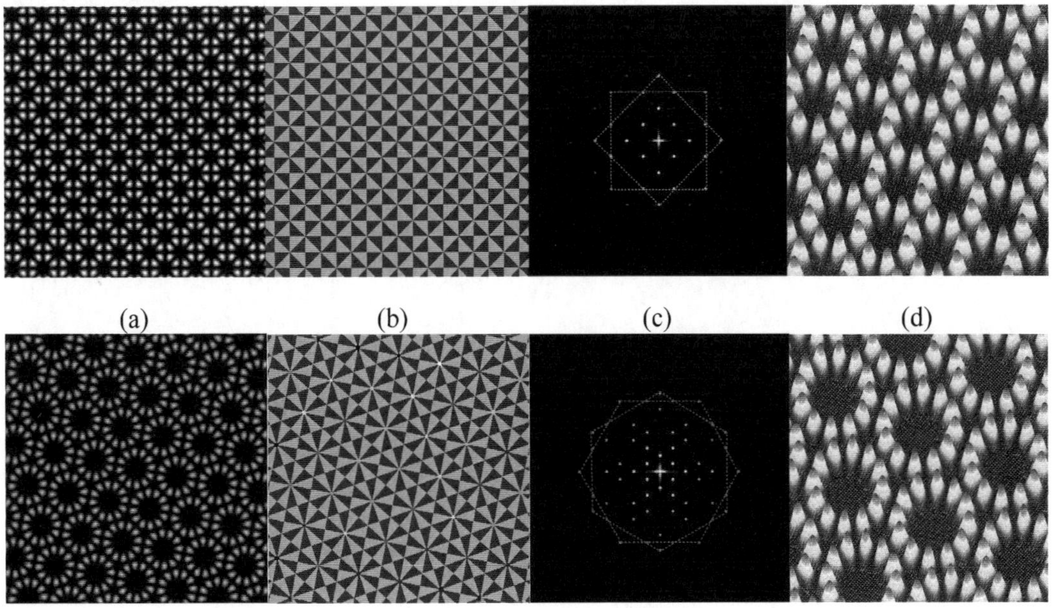

(a) (b) (c) (d)

(e) (f) (g) (h)

Figure 1. The simulation results. (a)-(d)Intensity pattern, Phase distribution, far-field pattern and three dimensional (3D) intensity pattern of regular octagon photonic lattice. (e)-(h)Intensity pattern, Phase distribution, far-field pattern and 3D intensity pattern of regular dodecagon photonic lattice.

Based on the Bessel beams and Fourier transformation [10], a series of results on the light field has been simulated. As is shown in Figure 1(a), the light field contains periodic distribution of a large number of regular octagons [11]. Also, the phase distribution in Figure 1(b) are arranged periodically. Comparing (a) and (b), there is a π-phase difference between two lattice sites [12]. From the far-field pattern (c), we can see that the distribution of far-field is a superposition of two equal squares. Also, the distribution in (g) is a superposition of two same hexagon. As the 3D intensity pattern shows, the intensity of regular octagon lattice is like ridges arranged periodically. It suggests the periodicity of regular octagon lattice and the non-diffraction of Bessel beams. According to the figure from (e) to (h), we can conclude that the phenomenon of regular dodecagon lattice is similar to that of regular octagon lattice.

3. Analyzation of the energy band structure

For further analyses, we calculated the band structure [13] of the two complicated photonic lattice. The photonic lattice has the great application value and the huge development potential due to the existence of energy band gap. Essentially, we need the master equations of 2D photonic lattice first:

$$\frac{1}{\varepsilon(r')}\left(\frac{\partial^2}{\partial x^2}+\frac{\partial^2}{\partial y^2}\right)E_z(r')+\frac{w^2}{c^2}E_z(r')=0 \qquad \text{(TE mode)} \tag{1}$$

$$\frac{\partial}{\partial x}[\frac{1}{\varepsilon(r')}\frac{\partial H_z}{\partial x}]+\frac{\partial}{\partial y}[\frac{1}{\varepsilon(r')}\frac{\partial H_z}{\partial y}]+\frac{w^2}{c^2}H_z(r')=0 \quad \text{(TM mode)} \tag{2}$$

In the TE mode, the electric field distribution is in XY plane. And we mainly studies band gap in this mode. In the TM mode, the magnetic field distribution is in XY plane. Next, we calculate the band structure by the PWE.

The Fourier series expansion of the periodic dielectric function in reciprocal space:

$$\frac{1}{\varepsilon(r')}=\sum_G k(G)e^{(iG\cdot r')} \tag{3}$$

And, $G=l_1b_1+l_2b_2$, which is basic vectors in reciprocal space. The $k(G)$ is Fourier coefficients of $\frac{1}{\varepsilon(r')}$.

Expanding the $E_z(r')$:

$$E_z(r')=\sum_G E_z(G)e^{[i(k+G)\cdot r']} \tag{4}$$

The eigenvalue equation in TE mode is deduced via substituting (3) and (4) into the master equation of 2D photonic lattice (1):

$$\sum_G k(G-G')|k+G||k+G'|E_z(G')=\frac{w^2}{c^2}E_z(G) \tag{5}$$

The expansion of $H_z(r')$:

$$H_z(r')=\sum_G H_z(G)e^{[i(k+G)\cdot r']} \tag{6}$$

Similarly,the eigenvalue equation in TM mode:

$$\sum_G k(G-G')(k+G)\cdot(k+G')H_z(G')=\frac{w^2}{c^2}H_z(G) \tag{7}$$

The energy band structure can be calculated by using the eigenvalue equations above [14,15].

IOP Conf. Series: Journal of Physics: Conf. Series **867** (2017) 012025 doi:10.1088/1742-6596/867/1/012025

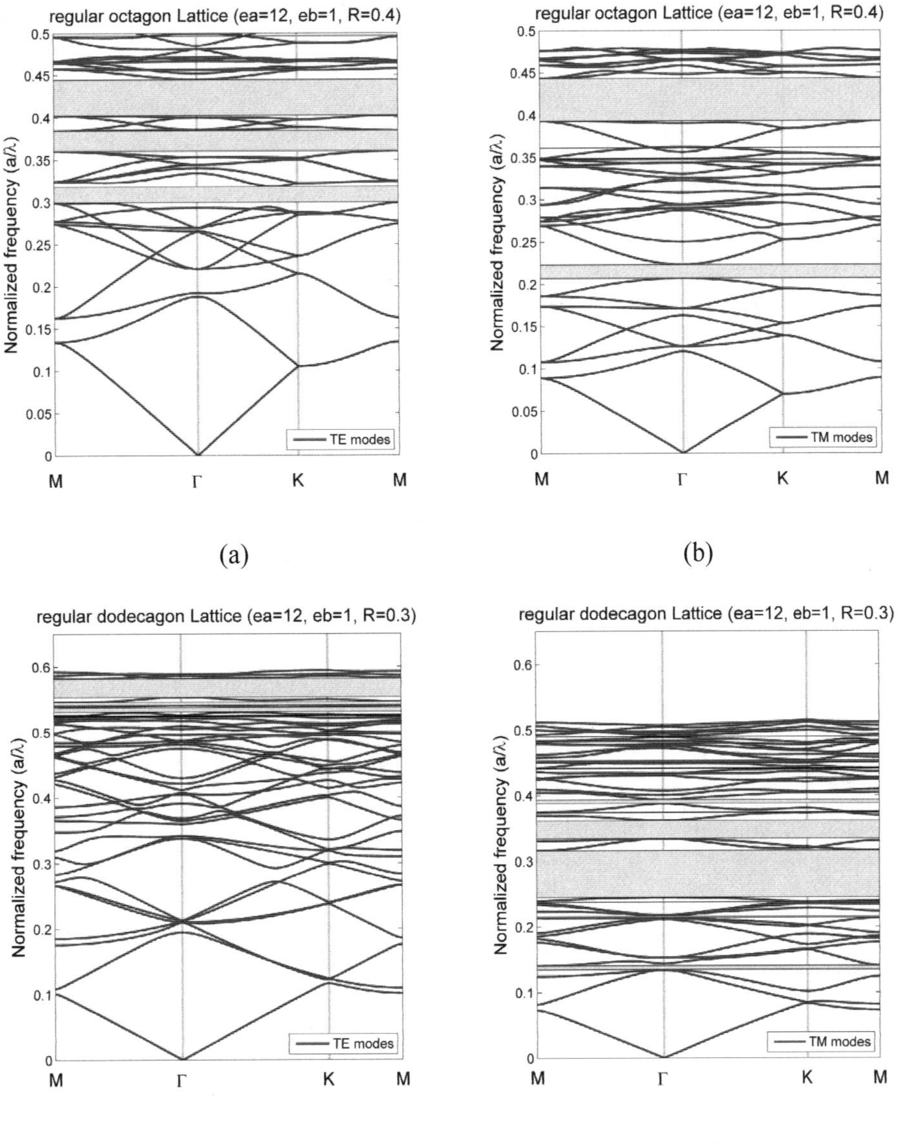

Figure 2. The simulation results of band structure. (a) and (b) The band structure of regular octagon lattice in TE mode and TM mode.(c) and (d) The band structure of regular dodecagon lattice in TE mode and TM mode.

Duo to the complicated structure of regular octagon lattice, we regard it as the compound lattice. The Fourier coefficients $k(G)$ of regular octagon lattice:

$$k(G) = \frac{1}{a_c}\int_{a_c} dr' \frac{1}{\varepsilon(r')} e^{(-iG\cdot r')} = \begin{cases} \frac{1}{\varepsilon_b} + [\frac{1}{\varepsilon_a} - \frac{1}{\varepsilon_b}]f \\ [\frac{1}{\varepsilon_a} - \frac{1}{\varepsilon_b}] 2 f_0 \frac{J_1(Gr_a)}{Gr_a} \sum_{n=1}^{4} 2cos(G \cdot P_n) \end{cases} \tag{8}$$

The f means the filling ratio of regular octagon lattice. ε_a and ε_b are the dielectric constants of lattice point and background medium respectively [11]. And the r_a is the radius of the lattice point. After the calculation, we got the P_n that are the coordinates of the lattice points in compound lattice: $P_1=(\frac{1+\sqrt{2}}{2},\frac{1}{2})=-P_{-1}$, $P_2=(\frac{1}{2},\frac{1+\sqrt{2}}{2})=-P_{-2}$, $P_3=(-\frac{1}{2},\frac{1+\sqrt{2}}{2})=-P_{-3}$ and $P_4=(-\frac{1+\sqrt{2}}{2},\frac{1}{2})=-P_{-4}$. And the unit is micron . Similarly, we got the Fourier coefficients $k(G)$ of regular dodecagon lattice:

$$k(G) = \frac{1}{a_c}\int_{a_c} dr' \frac{1}{\varepsilon(r')} e^{(-iG\cdot r')} = \begin{cases} \frac{1}{\varepsilon_b} + [\frac{1}{\varepsilon_a} - \frac{1}{\varepsilon_b}]f \\ [\frac{1}{\varepsilon_a} - \frac{1}{\varepsilon_b}] 2 f_0 \frac{J_1(Gr_a)}{Gr_a} \sum_{n=1}^{6} 2cos(G \cdot P_n) \end{cases} \tag{9}$$

The P_n in regular dodecagon lattice: $P_1=(\frac{1}{2},\sqrt{\frac{7}{4}+\sqrt{3}})=P_{-1}$, $P_2=(\sqrt{1+\frac{\sqrt{3}}{2}},\sqrt{1+\frac{\sqrt{3}}{2}})=P_{-2}$, $P_3=(\sqrt{\frac{7}{4}+\sqrt{3}},\frac{1}{2})=P_{-3}$, $P_4=(\sqrt{\frac{7}{4}+\sqrt{3}},-\frac{1}{2})=P_{-4}$, $P_5=(\sqrt{1+\frac{\sqrt{3}}{2}},-\sqrt{1+\frac{\sqrt{3}}{2}})=P_{-5}$ and $P_6=(\frac{1}{2},-\sqrt{\frac{7}{4}+\sqrt{3}})=P_{-6}$. As is shown in Figure 2(a) and (b), there are three band gaps in TE mode and two gaps in TM mode [15]. The band gap in regular octagon lattice is wide. More importantly, there is a large gap in the range of (0.402~0.445), whatever the mode is TE or TM, which means a complete band gap in 2D regular octagon lattice. The complete band gap [16] can achieve the light induction at any direction in range of two dimensional space. It portends that there is great application value for regular octagon photonic lattice. From Figure 2(c) and (d), there are also band gaps of regular dodecagon lattice in TE and TM mode. But, the gaps of it are not complete. Comparing the two lattices, the regular octagon lattice is better in application.

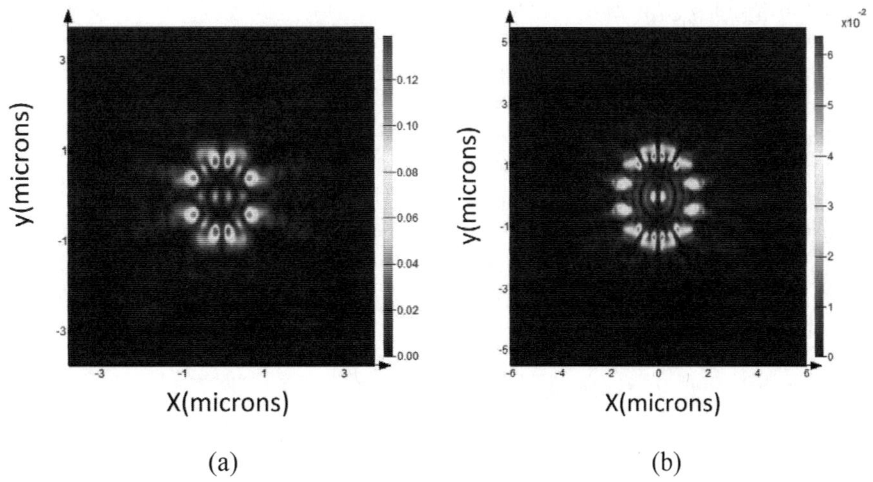

Figure 3. Transmission image. (a) and (b) The simulation results of output image on the back of the regular octagon lattice and the regular dodecagon lattice.

As Figure 3 shows, the transmission image was simulated by the method of FDTD. We can see the solitons arranged in the regular octagon or the regular dodecagon on the back when the Gauss beams irradiate on a cell lattice of the front surface in Figure 3(a) and (b) [17]. The localization wavepacket transition [18] results from band gaps of the complicated lattice [8].

4. Conclusion

The research mainly studies on the 2D complicated regular polygon photonic lattices, regular octagon lattice and regular dodecagon lattice. The lattice is a superposition of two equal simple lattices with a certain angle and phase. And the light field of the complicated lattice has been derived in this paper. The regular octagon lattice and regular octagon lattice possess the huge application value due to the energy band gaps. The regular octagon lattice which has the wide complete band gap. And we use the method of PWE to calculate the band structure. Moreover, the solitons can be observed in the photonic lattice we design [19]. This paper only analyses the complicated lattice preliminarily. Thus, many problems are still to be studied and solved.

Acknowledgments

This work is supported by the National Natural Science Foundation of China (Grant nos. 11304187, 11404196, 11574185,11404196 and 11604183). A Project of Shandong Province Higher Educational Science and Technology Program (Grant no. J14LA55) and China postdoctoral science foundation projects (Grant no.2015M582126).

References

[1] Shvedov V G et al. 2012 *Opt. Lett* **37** 1934-6
[2] Liu S et al.2012 *Opt. Lett* **37** 2184-6
[3] Zong Y et al. 2016 *Opt. Express* **24** 8877-85
[4] Zhang P et al. 2017 *Opt. Lett* **42** 915-8
[5] S. John et al. 1987 *Phys. Rev. Lett* **58** 2486
[6] B. Terhalle et al. 2011 *Phys. Rev. Lett* **106** 083902

IOP Conf. Series: Journal of Physics: Conf. Series **867** (2017) 012025 doi:10.1088/1742-6596/867/1/012025

[7] Y. Gao et al. 2014 *IEEE Photonics J* **6** 2201806
[8] Liu S et al. 2012 *Appl. Phys. Lett* **100** 061907
[9] Zhao J et al. 2015 *Sci. Bull* **60** 1157-69
[10] Gao Y et al. 2017 *Opt. Commun* **389** 123-7
[11] Freedman B et al. 2006 *Nature* **440** 1166-9
[12] Gao Y et al. 2014 *IEEE Photonics J* **6** 1-6
[13] Chremmos I D and Efremidis N K 2012 *Phys. Rev.* A **85** 063830
[14] Kuzmiak V, Maradudin A A and Pincemin F 1994 *Phys. Rev.* B **50** 16835
[15] Plihal M and Maradudin A A 1991 *Phys. Rev.* B **44** 8565
[16] Yang Y and Wang G P 2007 *Opt. Express* **5** 5991-6
[17] Ni P et al. 2016 *Opt. Express* **24** 2420-6
[18] Huang C et al. 2016 *Scientific reports* **6** 32546
[19] Xia S et al. 2015 *Opt. Express* **23** 4397-405

IOP Conf. Series: Journal of Physics: Conf. Series **867** (2017) 012026 doi:10.1088/1742-6596/867/1/012026

Pyroelectric generation of 2D spatial soliton sets in a bulk of lithium niobate crystal

V Ryabchenok, V Shandarov, and A Perin

Tomsk State University of Control Systems and Radioelectronics, Lenin Prospect 40, Tomsk, 634050, Russia

E-mail: vryabchenok@gmail.com

Abstract. The generation of two-dimensional bright spatial soliton sets in lithium niobate sample has been experimentally demonstrated at light wavelength of 532 nm, contribution of pyroelectric effect into nonlinear optical response of the crystal, and spatial modulation of one-dimensional beam along direction normal to the crystal optical axis. Diameters of soliton beams and channel waveguides formed within the crystal bulk by these solitons are near to 20 μm at light polarization corresponding to extraordinary wave of the crystal.

1. Introduction

Optical waveguide elements and waveguide structures with potential optical reshaping are important elements for photonic devices and systems. Such elements may be formed in electrooptic crystalline materials or layered structures exploiting the effects of spatial optical solitons. The one of most interesting electrooptic crystals promising for similar applications is lithium niobate ($LiNbO_3$). The nonlinear optical response of this crystal doped with some impurities like iron (Fe) or copper (Cu) is observed at very low light intensities [1 - 3]. However, this response is self-defocusing that allows formation only dark spatial solitons in this material at usual conditions. To overcome this limitation, the drift mechanism of charge carrier transport using at the photorefractive effect has been proposed to compensate large photovoltaic contribution into that transport [4]. This method is effective for crystals with low level of photorefractive impurities, but it requires high-voltage electric field applied to the crystal sample. Another approach to solve that problem with no external electric field is based on ferroelectric properties of $LiNbO_3$. Forming bright spatial solitons becomes possible when pyroelectric effect contributes to its nonlinear optical response [5 - 7]. Using this effect, one can form two-dimensional bright spatial solitons and, correspondingly, channel optical waveguides in $LiNbO_3$ crystals with low photorefractive sensitivity [5]. One-dimensional bright spatial solitons may form in $LiNbO_3$ due to its wavelength-dependent photovoltaic properties [8]. The main aim of our work is the study of generation of 2D spatial soliton sets in congruent $LiNbO_3$ bulk at contribution of pyroelectric effect into material nonlinear response using laser radiation with wavelength of 532 nm.

2. Experimental setup and experimental results

The schematic of experimental setup is shown in Figure 1. Laser beam (λ=532 nm) with polarization corresponding to extraordinary wave of crystal is focused with a cylindrical lens (CL) onto a surface of amplitude mask (AM) on a glass plate in order to illuminate it with elliptic light spot (see inset to Figure 1). The light field with periodic spatial modulation of its intensity along direction orthogonal to

IOP Conf. Series: Journal of Physics: Conf. Series **867** (2017) 012026 doi:10.1088/1742-6596/867/1/012026

crystal optical axis obtained after this mask is optically projected with spherical lens (L) onto the entrance surface of the crystal sample (LN), placed on a surface of Peltier heater (PE). The crystal temperature is measured with a non-contact infrared thermometer. Light field with smallest ellipse axis oriented parallel to crystal optical axis propagates within the crystal along its Y-axis through distance about 10 mm. The sample dimensions make up 4×10×4 mm^3 along crystalline X, Y, Z axes. The spherical lens (L) forms the imaging of entrance or exit surfaces of the sample onto the sensitive matrix of a laser beam analyzer (BA). The light patters from these surfaces are studied visually and saved as the files in a computer memory. These images are compared with those obtained in the same sample with no amplitude mask but at the same light power, polarization, and other experimental conditions.

2.1. Formation of sets of two-dimensional bright spatial solitons in a crystal bulk using spatially modulated light field

Transverse size of light field is increased during its propagation within the medium due to the natural light diffraction. The beam divergence may be compensated when self-focusing optical nonlinearity of the medium comes into play. It may be observed in photorefractive $LiNbO_3$ when photovoltaic mechanism of charge carrier transport is totally suppressed by another physical mechanism. To reverse the sign of $LiNbO_3$ optical nonlinearity, the contribution of that mechanism should exceed the photovoltaic current contribution into nonlinear optical response of $LiNbO_3$. As it has been mentioned, we can use for that purpose the pyroelectric properties of ferroelectric $LiNbO_3$.

To generate the set of 2D bright spatial solitons in $LiNbO_3$ sample, we use periodic spatial modulation of light intensity in direction normal to the optical axis direction in $LiNbO_3$. Some particular experimental results are illustrated by images in Figure 2. Image (2a) corresponds to the light pattern at the entrance surface of the sample and natural diffraction changes its width along crystal optical axis at the exit surface (Images 2b, 2c). The nonlinear diffraction increases light pattern size along that direction at the exit surface due to the photorefractive effect (Image 2d). To compensate both, nonlinear and linear light diffraction, we heat the crystal sample up to the temperatures (60 – 75)°C depending on light power which ranges from 1 to 10 mW. It is illustrated by image (2e) for light power 5 mW, light pattern dimensions of 20×1000 μm at the entrance surface, and sample temperature of 75°C. Corresponding intensity profiles of these patterns are shown in Figure 3.

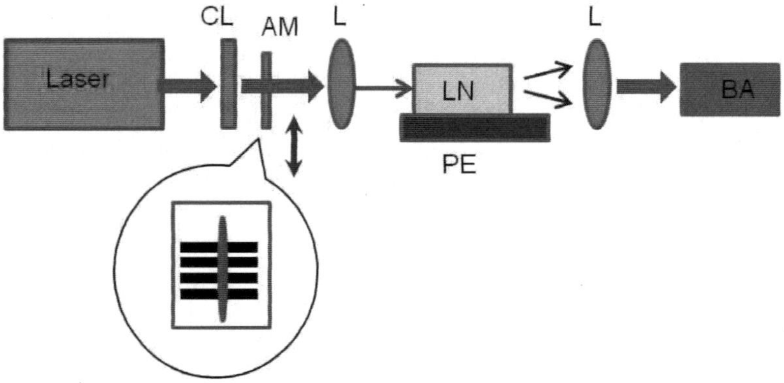

Figure 1. Experimental setup scheme. Inset demonstrates light field shape on AM surface.

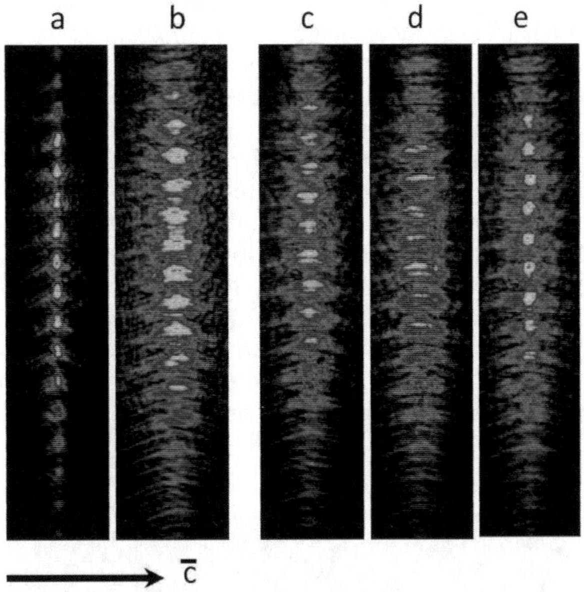

Figure 2. Light field images at the entrance surface of the crystal sample (2a), at its exit surface in linear regime (2b, c), in nonlinear regime (2d), and at the exit surface in conditions of total compensation of light field divergence along crystal optical axis (2e).

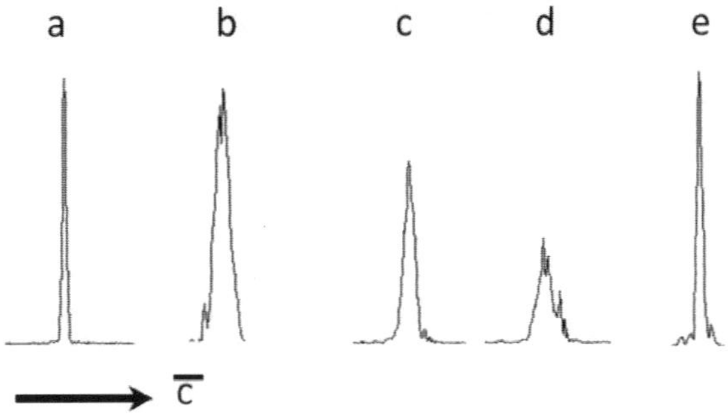

Figure 3. Intensity profiles of ight fields at the entrance (3a), and the exit surfaces of crystal sample (3 b – e). Letters correspond to situations, discussed in a capture to Figure 2.

These results clearly demonstrate formation of sets of 2D bright spatial solitons in the sample tested. It is due to the almost uniform crystal heating with Peltier element. The change of the sample temperature results in the change of spontaneous polarization P_S of ferroelectric $LiNbO_3$ sample that generates within it the pyroelectric field E_p [5]:

$$E_p = -\frac{1}{\varepsilon_0 \varepsilon_r}\frac{dP_S}{dT}\Delta T \qquad (1).$$

Here ε_0 and ε_r are the dielectric constants of vacuum and the crystal, and ΔT is a temperature variation in the illuminated part of material with respect to its initial temperature. By means of the crystal heating, the pyroelectric field E_p lowers the refractive indices of the whole sample due to the linear electrooptic effect on value of $\Delta n=0.5n^3 r_{\mathrm{eff}} E_{\mathrm{p}}$. Here n is average refractive index for the linearly polarized light wave, and r_{eff} is effective electrooptic constant. At the same time this pyroelectric field is screened in the illuminated area of the crystal sample due to the spatial redistribution of charge carriers existing in the illuminated area because of the photovoltaic current via light influence. The experiments show that the steady-state formation of regularly separated 2D bright spatial solitons is generated when light beam is spatially modulated.

We also study formation of 1D bright spatial soliton in the same crystal sample with light beam of Gaussian shape along direction normal to the optical axis. In that case the laser beam is directly focused onto the entrance surface of the sample. Some particular results of that experiment are shown in Figure 4. Here we also observe the linear diffraction of shit-shaped light beam at its propagation from the sample entrance (Image 4a) to its exit surface (Image 4b), the nonlinear contribution to the light diffraction due to the photorefractive effect (Image 4c), and the diffraction compensation by the pyroelectric effect (Image 4d) at the sample heating. However in the last image we observe separation of this shit-shaped light beam along its largest dimension to the almost regular structure of 2D spots. These spots correspond to separate 2D bright spatial solitons formed from 1D beam due to the spatial modulation instability [9 - 11]. It should be noted that this configuration of spatial soliton formation is very sensitive to the variations of experimental parameters. Using the spatially modulated light pattern for the spatial soliton excitation allows control of distances between separate 2D bright solitons.

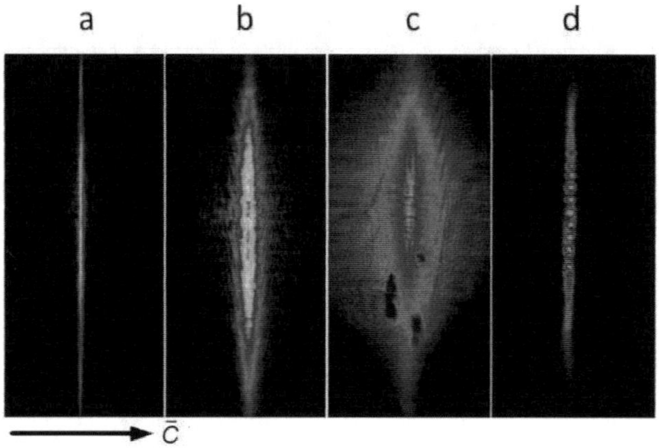

Figure 4. Light field images at the entrance surface of the crystal sample (a), its exit surface in linear regime (b), at exit surface and photorefractive divergence (c), and at exit surface in conditions of beam separation on 2D spots due to the spatial modulation instability (d).

3. Conclusion

In conclusion, our experimental results demonstrate the possible generation of sets of 2D bright spatial solitons in $LiNbO_3$ samples with predicted distances between them due to the contribution of pyroelectric effect into its nonlinear optical response and spatial optical modulation of 1D light beam along direction normal to the crystal optical axis.

Acknowledgments

This study was carried out with the financial support of Ministry of Education and Science of Russia (within the task N 3.1110.2017/PCh of the project part).

References

[1] Kip D 1998 *Appl. Phys. B* **67** 131
[2] Petrov M P, Stepanov S I and Khomenko A V 1991 *Photorefractive Crystals in Coherent Optical Systems* (Berlin: Springer-Verlag)
[3] Chen Z, Segev M, and Christodoulides D 2012 *Rep. Prog. Phys.* 2012 **75** 086401
[4] Fazio E, Renzi F, Rinaldi R, Bertolotti M, Chauvet M, Ramadan W, Petris A, and Vlad V 2004 *Appl. Phys. Lett.* **85** 2193
[5] Safioui J, Devaux F, and Chauvet M 2009 *Optics Express* **17** 205

[6] Popescu S, Petris A, Vlad V 2013 *J. Appl. Phys.* **113** 213110
[7] Perin A, Shandarov V, Ryabchenok V 2016 *Phys. of Wave Phenom.* **24** 7
[8] Zhang M, Huo G, Zhang Y, Kang Y, and Duan Z 2012 *Laser Physics* **22** 1295
[9] Mamaev A, Saffman M, Anderson D, and Zozulya A 1996 *Phys. Rev. A.* **54** 870
[10] Królikowski W, Luther-Davies , Denz C 2003 *IEEE Journ. Of Quant. Electron.* **39** 3
[11] Fang H, Malendevich R, Schiek R, and Stegeman G 2000 *Optics Letters* **25** 1786

IOP Conf. Series: Journal of Physics: Conf. Series **867** (2017) 012027 doi:10.1088/1742-6596/867/1/012027

Dynamic Holographic Interferometry with Doped $Sn_2P_2S_6$ Photorefractive Crystals

A Grabar[1], P Mathey[2], M Tsyhyka[1], G Gadrét[2], and I Stoika[1]

[1] Institute of Solid State Physics and Chemistry, Uzhgorod National University, Pidhirna st. 46, 88000 Uzhgorod, Ukraine
[2] Laboratoire Interdisciplinaire Carnot de Bourgogne (ICB), UMR 6303 CNRS - Université de Bourgogne Franche-Comté, 9 Av. A. Savary, BP 47 870, F-21078 Dijon Cedex, France

al_grabar@hotmail.com

Abstract. We present results obtained with several schemes of dynamic interferometers based on the photorefractive parameters of Sb-doped $Sn_2P_2S_6$ crystals. Besides the typical two-wave mixing scheme, we realized a lensless dynamic interferometer and a single-beam interferometer exploiting the beam fanning. The frequency range for all these schemes can be enlarged with use of heterodyne principle that also allows to measure all the parameters of the testing vibrations. We show that the phase-modulating technique can be used for testing the material's photorefractive parameters.

1. Introduction

Among the most perspective applications of photorefractive crystals is dynamic interferometry based on two-wave mixing of laser beams that can be used for remote detection of small dynamic phase variations produced by vibrations of various objects [1]. The main advantage of such schemes is their adaptation to slow variation of phase, i.e. occurring in time longer than the response time of the photorefractive grating. The basic element of the interferometer is the photorefractive crystal with fast response and high two-wave mixing gain at the used laser wavelength. We use a $Sn_2P_2S_6$ single crystal doped with Sb (1 mol. %) which is characterized by diffusion mechanism of the photorefractive grating formation, and demonstrates high values of the two-wave mixing gain (up to 18 cm^{-1} at 633 nm light wavelength) along with a relatively short response time (10...100 ms) [2].

2. Experimental schemes of the dynamic interferometers

In our works we examined several variants of interferometric schemes constructed on the base of $Sn_2P_2S_6$:Sb crystal. The first one is the conventional two-beam scheme (Fig. 1) that allows to detect the parameters of vibrations of one mirror driven by a piezoelectric actuator and reflecting the probe laser beam. The scheme acts as a frequency filter in which the cut-off frequency depends on the laser beam intensity and photorefractive grating spacing. Under our experimental conditions, this value is in a range of 10 Hz that corresponds approximately to the inverse time constant of the photorefractive grating formation (Fig. 2). In order to test a wider frequency range of vibrations, we used the heterodyne principle where the reference beam is also phase-modulated. Under these conditions, the interference pattern inside the photorefractive crystal, that forms the dynamic hologram, vibrates with a frequency

IOP Conf. Series: Journal of Physics: Conf. Series **867** (2017) 012027 doi:10.1088/1742-6596/867/1/012027

equal to the difference between the frequencies of the signal (f_S) and reference (f_P) beam phase modulations. By linear sweeping the frequency of the reference beam phase modulation f_P in some range around f_S one can detect the object vibration frequency with a high precision. As shown in Fig. 3, the measured intensity of the amplified signal beam, detected with variation of the f_P, is symmetrical relative to the modulation frequency equal for the both interfering beams, and at equalized frequencies ($f_S = f_P$) depends on the relative phase of the vibrations, as illustrated by curves 1 and 2 in Fig. 3. Other curves correspond to intermediate values of the phase difference. Using this scheme the amplitude and relative phase of the testing point on the vibrating object can be evaluated.

The shape of the signal depends on the amplitude of the vibrating object, and higher harmonics appear in the signal at higher amplitudes of the "signal" (or "pump") phase modulation. This gives a possibility to measure the amplitude of the vibration of the objects by an analysis of the waveform of the signal at various amplitudes of the "pump". The relative phases of the modulations of the "signal" and "pump" also can be estimated from the signal level at equalized modulation frequencies. Partially, the sweeping of the "pump" modulation frequency allows to detect the vibration frequency of the object with a high precision.

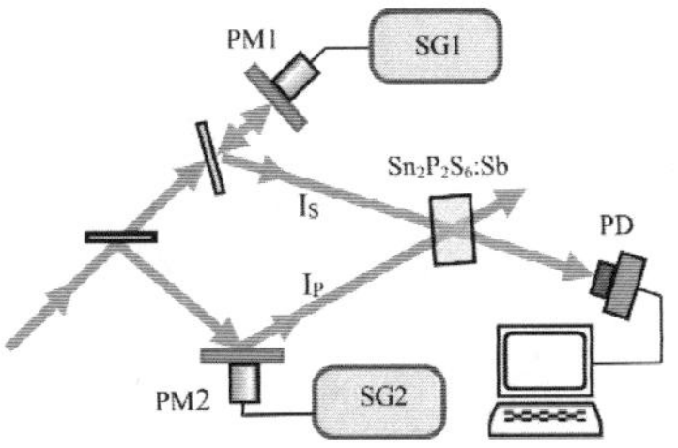

Fig. 1. The optical scheme of the heterodyne dynamic interferometer. Here PM1, PM2 are piezo-mirrors, SG1 and SG2 are sound generators, PD is photodiode, I_P and I_S are interacting beams of He-Ne laser (633 nm).

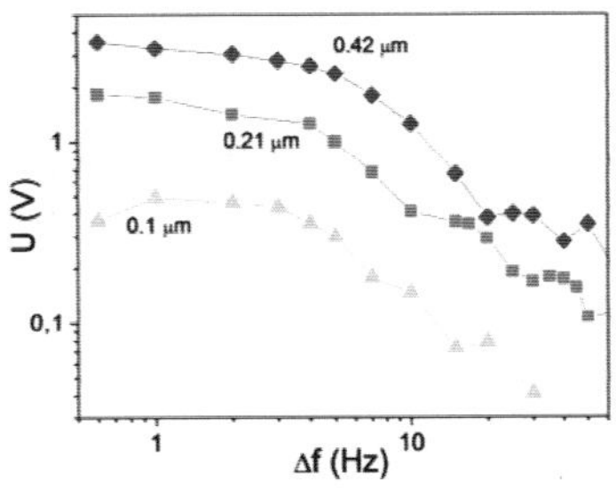

Fig. 2. The dependences of the amplitudes of the signal beam intensity as a function of the frequency detuning obtained at various amplitudes of the vibration of PM1 and PM2.

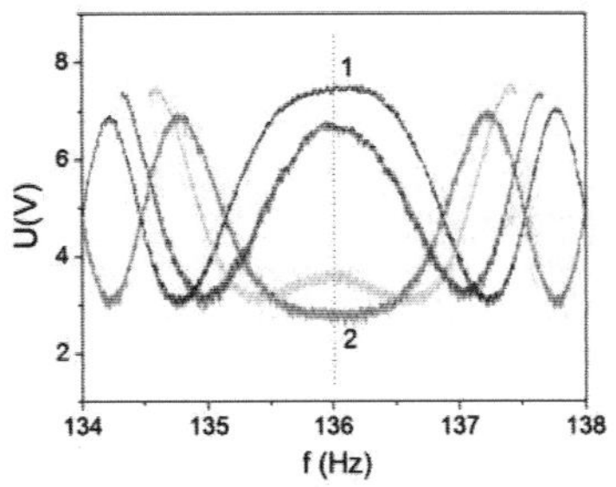

Fig. 3. Variation of the signal beam intensity I_S at linear in time variation of f_P when object (PM1) is vibrating with frequency f_S = 136 Hz. Curves 1 and 2 correspond to “in-phase” and “anti-phase” vibrations, when $f_S = f_P$.

The high gain value in the $Sn_2P_2S_6$:Sb crystal allows also to construct a lensless scheme applicable for optical detection of micro-vibrations, partially to put in evidence 2D-patterns of vibrating surfaces illuminated by a laser beam. This scheme is based on the image formation in a space-limited amplifying dynamic hologram, formed by a narrow (focused) reference beam, and which acts similarly to an optical pinhole in the known "camera obscura" [3]. We demonstrate that this scheme can be used to put in evidence the dynamical wavefront distortions of the coherent light scattered by a vibrating rough surface. In our experiments this is illustrated by the real-time observation of the standing vibration of a light-scattering object [4]. We also show that this lensless scheme can be used for the optical detection of microvibrations whose frequency is much higher than the reversal response time of the photorefractive material by using a heterodyne principle, like in the above two-wave mixing scheme. It allows also to measure the parameters of the vibrations in different points of the scattering object by using the same procedure (frequency sweeping with further amplitude and phase variation of reference beam). The advantages of this scheme in comparison with the known dynamical interferometers exploiting the two-wave mixing effects are the following: this is a focus-free scheme which does not require the adjustment of the interacting beams and focusing elements, it is able to get the dynamic picture from a rough light-scattering object. With this scheme one can detect the phase modulation in 2D, or one can detect simultaneously the vibrations of several points on the object.

Basing on the strong and fast beam fanning in $Sn_2P_2S_6$:Sb [5] one can construct a one-beam interferometer, shown in Fig. 4. The main advantage of this scheme is that it does not require to equalize the optical arms, and the phase modulation of the input beam can be detected by measuring the scattered intensity. By introducing the reference phase modulation (i.e. the second piezo-driven mirror) into this scheme, one can benefit from the advantages of the heterodyne principle and measure the frequency, amplitude and phase of the vibrating object by compensation of the phase modulation in a way similar to that described above, i.e. by tuning the frequency, amplitude and phase of the vibrations of the second mirror. As an example, the typical oscillogram of the fanning intensity, measured at simultaneous modulation of the beam by two sequent mirrors with different frequencies, is shown in Fig. 5. By varying the frequency of PM2 vibrations, one can reach equal amplitudes and frequencies, that are well seen on the Fourier-spectrum of the detected vibrations. We can also reach phase compensation by varying the phase of PM2 vibration.

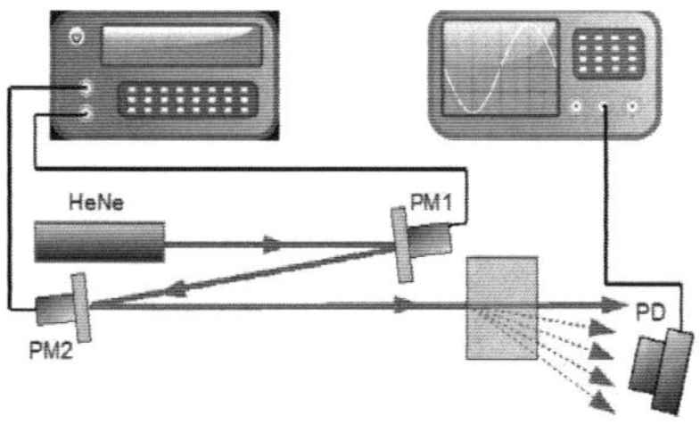

Fig. 4. The scheme of the one-beam dynamic interferometer based on the beam fanning. PM1 and PM2 are piezomirrors, driven by a sound generator.

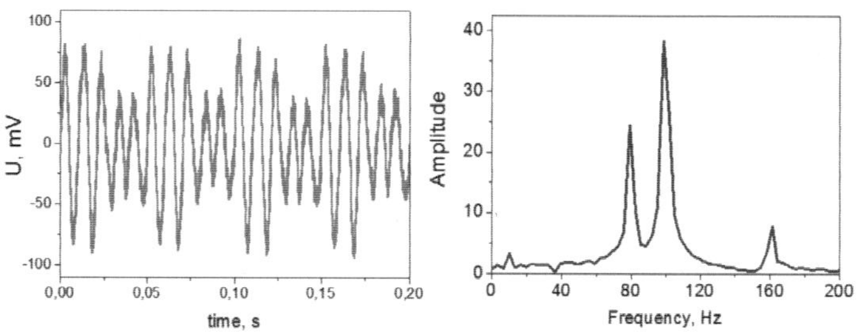

Fig. 5. The typical oscillogram of the beam fanning intensity and its Fourier-transform, measured in the one-beam scheme shown in Fig. 4, at simultaneous vibrations of PM1 and PM2 with frequencies of 80 Hz and 100 Hz respectively.

3. Conclusions

Thus, the Sb-doped $Sn_2P_2S_6$ photorefractive crystal allows to construct several schemes of the dynamic interferometers. Besides, the phase-modulation technique can be used for measurements the material parameters, partially to study the peculiarities of the two-wave mixing gain dispersion in various doped compounds of the $Sn_2P_2S_6$ crystals.

References

[1] Kamshilin A, Romashko R and Kulchin Yu 2009 *J. Appl. Phys.* **105** 031101

[2] Kedyk I, Mathey P, Gadrét G, Grabar A, Fedyo K, Stoika I, Prits I and Vysochanskii Yu 2008 *Applied Phys.B* **92** 549

[3] Grabar A, Mathey P, Iegorov R and Gadrét G 2011 *Optics Commun.* **284** 5361

[4] Grabar A, Mathey P, Gadrét G, Tsyhyka M and Stoika I 2015 *Abstracts of PR'15 - "Photorefractive photonics" (Villars, Switzerland)*

[5] Grabar A, Mathey P and Gadrét G 2014 *Journ. Opt. Soc. Amer. B* **31** 980

IOP Conf. Series: Journal of Physics: Conf. Series **867** (2017) 012028 doi:10.1088/1742-6596/867/1/012028

Index and gain gratings in Nd-YVO4 - Applications to speckle vibrometry and photoacoustic detection

Baptiste Jayet, Jean-Pierre Huignard, François Ramaz

Institut Langevin, ESPCI ParisTech, PSL Research University, CNRS UMR7587, INSERM U979 UPMC 1 rue Jussieu, 75005 Paris, France

Corresponding author: francois.ramaz@espci.fr

Abstract. We perform wave mixing and dynamic holography by exploiting gain saturation in diode pumped laser media like Nd-YVO4. Such crystals offer the possibility of performing adaptive gain interferometers with a sub ms response time, typically between 100- 200μs. This temporal response is of great interest for applications such as speckle interferometry and acousto-optic detection for tissue imaging in biophotonics. In this work we operate in new conditions of high sensitivity with a linear response due to a refractive index contribution in the gain media. It originates from the wavelength mismatch of about 0.1 nm between the incident Nd-YAG probe laser and laser transition in diode pumped Nd-YVO4 amplifier. This permits the recording of two types of volume holograms respectively a gain and a phase volume holograms. It is the index grating which provides a linear response to incident low phase modulation at frequencies higher than 10 kHz. Experimental results confirm this behavior for adaptive interferometry and speckle vibrometry. In addition, the fast response makes the two wave mixing interaction in Nd-YVO4 very interesting for the optical detection of large bandwith photoacoustic waves excited by a pulsed ns Q-Switch laser at 532 nm.Temporal photoacoustic signals and images of small objects behind a scattering media will be shown.

1. Wave mixing and vibrometry in gain laser media

When performing interferometry on a scattering surface such as biological media, the use of a conventional interferometer results in sensitivity mainly limited to the SNR of one speckle grain. A convenient solution to compensate for the low spatial coherence of the complex signal wavefront is to introduce a dynamic wavefront adaption stage into the interferometer. This is usually done using holographic two wave mixing (2WM) in a nonlinear crystal inducing the beam coupling effects between the incident interfering signal and reference waves. It thus results in perfect spatial matching of the signal and self-diffracted waves leading to a very sensitive linear response of the interferometer with respect to phase changes of the signal wave [1]. An original approach to perform wave mixing and wavefront adaption consists in exploiting gain saturation in laser media. Such media offers the possibility of dynamic holography with laser gain and with a sub-ms response time, typically 100-200 μs [2-4]. These performances are superior to photorefractive crystals in particular in term of speed. The experimental setup used is shown in figure1. It consists of an incident 1W CW Nd-YAG at 1064 nm illuminating the sample under test and the 2WM is done by interfering the signal and plane wave reference wave into the Nd-YVO4 gain laser media which size is 5x5x5 mm. The crystal is diode pumped at 808 nm by a high power diode able of delivering 100W quasi CW-10ms pulse at 10Hz. The holographic gain region is about 300μm diameter and 5mm long. The crystal is anti-reflected coated at

808 and 1064 nm. The interference fringes create a volume hologram due to the gain saturation, ie the dependence of the gain versus the fringe intensity. An important characteristic of the experiment relies on the fact that we record both an amplitude and a phase volume gain grating. The model of 2WM developed in [3-5] takes account of this index spatial modulation which can be decomposed into the sum of two terms $\beta nr + \beta r$ respectively linked to the non resonant and the resonant contributions. Considering our experimental conditions it will be shown that the wavelength mismatch of 0.1 nm between the gain curves of Nd-YAG and of the Nd-YVO4 is responsible for the index grating (Kramers-Kronig relations) [5]. Usually this additional phase modulation is neglected but recent interferometric experiments may efficiently exploit the index contribution in laser gain media. As this refractive index modulation is local with respect to the position of the incident fringes it directly yields a linear response of the dynamic interferometer with a high sensitivity. In other terms, the detection is linear with respect to the small amplitude of vibration of the object structure illuminated by the Nd-YAG laser. We outline that in case of a purely amplitude gain grating the response of the interferometer would be quadratic. The two types of responses, linear and quadratic have been confirmed in experiments with vibrating frequencies higher than 10 kHz [3]. In this experiment due to the refractive index grating component the amplitude of displacement is of few A°.

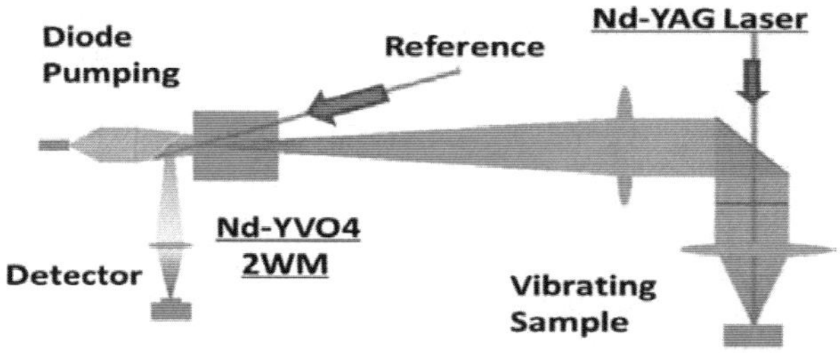

Fig 1. 2WM with diode pumped Nd-YVO4 laser gain media and applications to interferometry.

2. Application to Photoacoustic imaging

We also propose to apply this type of adaptive interferometer to probe the vibrations generated by a photoacoustic effect. In these experiments the acoustic wave is generated by local absorption of a short 4 ns laser pulse. This technique is well developed for 3D imaging in biological tissues where detection of the acoustic wave is made by conventional piezoelectric transducers. The presented all optical method has the potential advantage of realizing a non-contact detection with a much larger bandwith.The setup used achieve optical detection with the Nd-YVO4 adaptive interferometer [3]. The photoacoustic effect is generated using a pulsed 4ns - Q Switch 532 nm Nd-YAG laser delivering 150 mJ per pulse. When probing the surface of the sample with the adaptive interferometer it is possible to reconstruct the acoustic wave front and to image the object emitting the acoustic wave. Experimental results will be shown when using a small object of diameter 170 μm placed behind a clear gel of agar and a piece of cooked chicken breast to mimic a weakly scattering media as encountered in biophotonics applications. We demonstrate the possibility of using an adaptive holographic laser gain interferometer to probe acoustic waves generated by photoacoustic effect. Its sensitivity benefits from the gain grating index component due to the wavelength shift between the gain curves of Nd-YAG and Nd-YVO4. Moreover the fast response time of the interferometer (less than 100μs) enables to compensate parasitic vibrations (frequency <10 kHz) while being able to detect high frequency photoacoustic signals in the MHz range. These novel features open very significant characteristics for applications in nondestructive testing or in vivo biophotonics imaging.

IOP Conf. Series: Journal of Physics: Conf. Series **867** (2017) 012028 doi:10.1088/1742-6596/867/1/012028

References

[1] Kamshilin A A, Romashko R V, Kulchin Y N 2009 J. Appl. Phys. 105 031101
[2] Damzen M J, Boyle A, Minassian A 2005 Opt. Lett. 30 2230
[3] Jayet B, Huignard J P, Ramaz F. 2017 Opt. Lett. 42 2017
[4] Jayet B, Huignard J P, Ramaz R. 2014 Opt. Express 22 20622
[5] Soulard R, Brignon A, Huignard J P, Moncorge R 2010 J. Opt. Soc. Am. B 27 2203

IOP Conf. Series: Journal of Physics: Conf. Series **867** (2017) 012029 doi:10.1088/1742-6596/867/1/012029

Phase modulation detection with liquid crystal light valve and digital holography

Stefania Residori[1], Umberto Bortolozzo[1], Arnaud Peigné[2], Stephanie Molin[3], Daniel Dolfi[3], Jean-Pierre Huignard[4]

[1]Institut de Physique de Nice, UMR7010, CNRS, Université Nice-Sophia Antipolis, 1361 route des Lucioles, 06560 Valbonne - Sophia Antipolis, France
[2]Thales Underwater Systems, 525 route des Dolines, Sophia - Antipolis, France
[3]Thales Reasearch andTechnology France, 1 avenue Augustin Fresnel, 91767 Palaiseau, France
[4]Jphopto, 20 rue de Campo Formio, F-75013 Paris, France

E-mail: `stefaniaresidori@yahoo.com`

Abstract. Self-adaptive interferometry allows the measurement of small optical phase modulations in noisy environments and with strongly distorted optical wavefronts. We review two techniques of self-adaptive interferometers based on liquid crystals spatial light modulators, one obtained by using an optically addressed light valve, the second one realized by adopting a digital holography CMOS-LCOS scheme. The adaptive character of these two types of holographic interferometers will be analyzed and compared.

1. Introduction

Self-adaptive interferometry is a method for phase detection based on recording of dynamical holograms in a nonlinear medium [1], usually occurring via two-wave mixing processes [2]. A frequency bandwidth is dictated by the response time of the medium and provides a filtering function allowing the rejection of low frequency noise. Moreover, the adaptive character of the two-beam coupling allows maintains the quadrature condition (linear detection) without the need of feedback electronics. For these reasons adaptive holographic interferometers (AHI) provide attractive properties for phase demodulation with simplified architectures. Among the various applications, we can mention phase demodulators, acousto-optical imaging [3], increased sensitivity interferometers [4, 5, 6] and vibrometry at picometer scale [7, 8].

2. Phase detection with optically addressed light-valve

An AHI setup has been realized with an optically addressed spatial light modulator (OASLM), also called called liquid crystal light valves (LCLV), operating as the nonlinear medium [9]. The setup consists of a MachZehnder configuration in which the OASLM is the recombining medium (Fig.1). The interferometer behaves as an optical high-pass filter (cut-off of approximately 10 Hz), thus filtering slow phase disturbances and keeping the detection linear without the need of heterodyne or active stabilization. A sensitivity of 200 $\mu rad/\sqrt{Hz}$ has been estimated for normal laboratory conditions and without optimization of the laser source.

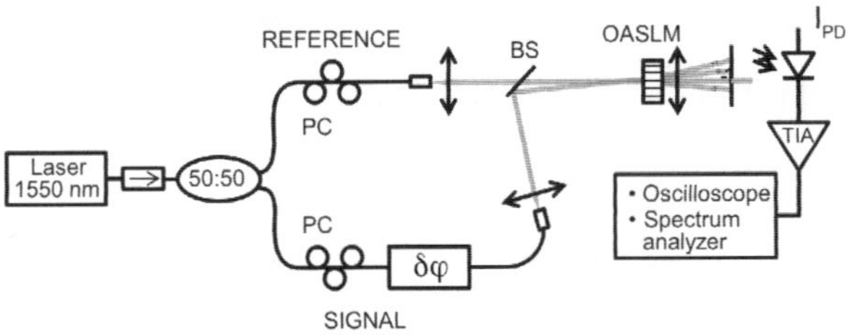

Figure 1. Adaptive holographic interferometer with optically addressed OASLM. A phase modulation $\delta\varphi$ is inserted in the signal arm.

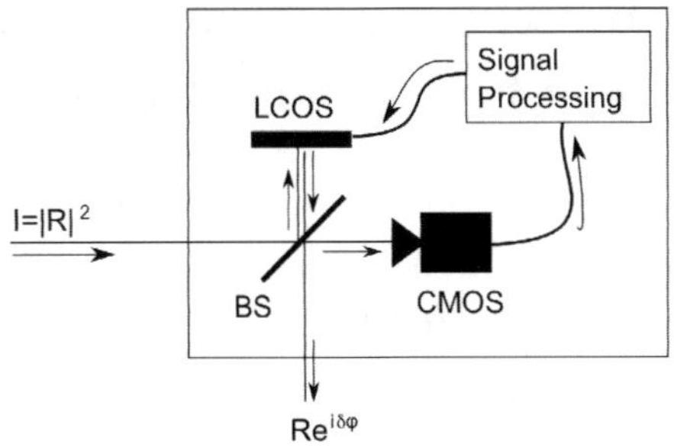

Figure 2. Digital holography setup.

3. Phase detection with a digital holography CMOS-LCOS setup

A digital off-axis holographic structure has been realized by employing a CMOS-LCOS (Liquid Crystal on Silicon) setup (Fig.2). The intensity fringe pattern formed by the two interacting waves is recorded by the CMOS matrix while the corresponding off-axis phase hologram is displayed on the LCOS SLM, from which the two incident waves self-diffract . The two-beam coupling gain is optimized through an electronic feedback, while the dynamic character of the hologram allows reaching a high sensitivity of the interferometric measurements even in unstable environments and with strongly distorted wavefronts [10].

4. Conclusions

Adaptive holographic interferometry are realized by performing two-wave mixing experiments in liquid crystal media, as analog OASLM or digital CMOS-LCOS architecture. These AHI systems are able to detect phase modulations with low amplitudes and with complex beams.

Acknowledgements

This work is supported by the Direction Générale de l'Armement, DGA, contracts ANR-11-ASTR- 0012, Astrid MEDUSE and ANR-14-ASMA-0004-01, Astrid Maturation HYDRE.

5. References

[1] Kamshilin A A, Romashko R V and Kulchin Y N 2009 *J. Appl. Phys.* **105** 031101
[2] Gunter P and Huignard J P 2006 *Photorefractive Materials and their Applications 1, 2 and 3* (Springer: New York, NY, USA)
[3] Benoit a la Guillaume E, Bortolozzo U, Huignard J P, Residori S. and Ramaz F *Opt. Lett.* **38** 287
[4] Bortolozzo U, Residori S and Huignard J P 2010 *Opt. Lett.* **35** 2076
[5] Bortolozzo U, Residori S, Rubin J and Huignard J P *Opt. Lett.* **36** 520
[6] Bortolozzo U, Residori S and J.C. Howell 2013 *Opt. Lett.* **38** 3107
[7] Bortolozzo U, Residori S and Huignard, J P 2009 *Opt. Lett.* **34** 2006
[8] Residori S, Bortolozzo U, Huignard J P and Kamshilin A A 2014 *Picometer Detection by Adaptive Holographic Interferometry*, in *Fundamentals of Picoscience*, Ed. by Sattler D K (CRC Press, Taylor & Francis, New York)
[9] Peigné A, Bortolozzo U, Residori S, Molin S, Nouchi P, Dolfi D and Huignard J P 2015 *Opt. Lett.* **40** 5482
[10] Bortolozzo U, Dolfi D, Huignard J P, Molin S, Peigné A and Residori S 2015 , *Opt. Lett.* **40** 1302

IOP Conf. Series: Journal of Physics: Conf. Series **867** (2017) 012030 doi:10.1088/1742-6596/867/1/012030

Optoelectronic tweezers based on photorefractive space charge fields: recent achievements and challenges

J F Muñoz-Martínez[1,2], A Alcazar[2], I Elvira[1], J Ramiro[2], A García-Cabañes[1], L Arizmendi[1] and M Carrascosa[1]

[1]Dep. de Física de Materiales, Universidad Autónoma de Madrid, 28049 Madrid, Spain
[2]Dep. de Mecánica de Fluidos y Propulsión Aeroespacial, Universidad Politécnica de Madrid, 28040 Madrid, Spain

Abstract: We present an overview of the operation of optoelectronic tweezers based on the photorefractive effect, paying special attention to the more recent results achieved by our group. The main challenges faced by the technique to enhance its technological potential are also discussed.

m.carrascosa@uam.es

1. Introduction

In the last decade a new kind of applications for photorefractive (PR) crystals, and particularly for those with high bulk photovoltaic effect, has emerged in the field of nano- and bio-photonics. It consists in using them as substrates to generate light-induced electric fields able to manipulate, trap and pattern micro- and nano-objects. The technique is usually called photovoltaic or photorefractive tweezers (PVT or PRT). First developments typically used $LiNbO_3$:Fe crystals with the usual orientation, having the polar axis parallel to the surface substrate (*x*- or *y*-cut), to get 1D particle patterns [1]. Initially, the structures had periods in the range of 50-500 μm that later became smaller, up to a few microns. 2D patterns were also attempted, but serious patterning limitations were found due to crystal anisotropy.

In this work, we present very recent achievements that remarkably contribute to develop the technique and its potential applications. We will also briefly discuss several challenges to further enhance the potential of the technique that is expected to become a leading kind of optoelectronic tweezers.

2. Recent results and discussion

2.1 Optimization of 2D patterns

First reported 2D particle patterns using *x*- or *y*-cuts presented the limitation that patterning along directions parallel to the *c*-axis were not possible because charge transport is parallel to the polar axis. An outstanding progress has been to use *z*-cut substrates, an unusual configuration for other PR applications. First work using *z*-cut was reported by Esseling et al. [2]. They used charged particles because they assumed that dielectrophoretic (DEP) forces acting on neutral particles are very week in this configuration [3]. Nevertheless, using highly doped crystals our group has demonstrated excellent performance for 2D trapping with neutral inorganic micro- and nanoparticles (NP) [3] and biological objects [4]. The experimental results have been accompanied with the development of an appropriate theoretical framework for this configuration [5] that satisfactorily explain the experimental patterns.

2.2 Action on charged and comparison with neutral particle operation

Although PVT are, in principle, able to manipulate both charged and neutral objects via electrophoretic (EP) or DEP forces, respectively, most initial experiments and theoretical calculations dealt only with neutral 1D particle patterning. To expand the possibilities of the technique we have recently investigated charged particle organization in *z*-cut and compared the differences between EP and DEP trapping by theory and experiments. A series of 2D nanoparticle patterns with charged and neutral aluminum NP on top of $LiNbO_3$:Fe crystals have been carried out. From these results we have obtained a further confirmation of three key aspects in this field: *i)* The possibility of successful trapping of charged particles, *ii)* the achievement of good quality 2D patterns, and *iii)* the ability of the developed theory to explain the main EP and DEP observed phenomenology.

2.3 Applications

Development of specific applications such as fabrication of metal NP platforms for plasmonic fluorescence (FL) enhancement from organic/biological molecules [6], patterning of micro- or nanometric biological species [4,7] or fabrication of photonic devices (fresnel lens and diffraction gratings) [8] have been recently reported. Another interesting kind of applications of the PV fields is the effective manipulation of micro and nano-droplets recently reported in several papers [9-12]. Let us describe two illustrative examples of PVT applications. First, PVT allow flexible patterning of metal NPs. In Figures 1a and 1b two patterns of Ag NP (25 nm diameter) are shown. After fabrication, and once the PR field has been erased, the samples are covered with a luminescent dye (DR1). Micro-luminescence measurements display in Figures 1c and 1d show a clear plasmonic enhancement (about a factor 10) of the dye FL just in the regions where the NP are located. Similar results have been obtained for fluorescein labeled DNA molecules on the metallic structures. In turn, in Figure 2, some examples of nanometric bio-objects patterning are shown.

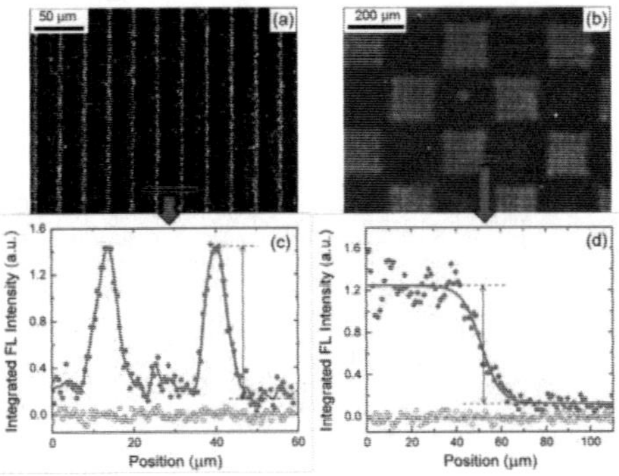

Figure 1: Fluorescence microscope photographs of (a) a 1D fringe patterns and (b) 2D square pattern. (c) and (d) show the corresponding micro-luminescence profiles of a region of the patterns.

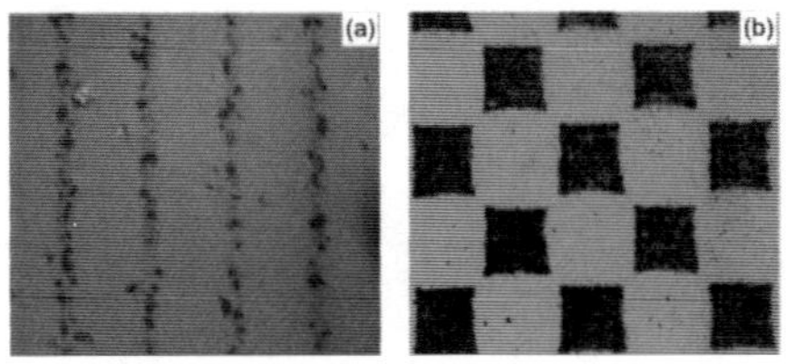

Figure 2: Optical microscope images of 1D fringes (Λ=14 µm) (a) and 2D square (side 200 µm) (b) patterns of nanometric spore and pollen fragments on *x*-cut and *z*-cut $LiNbO_3$:Fe crystals, respectively.

3. Overview and remaining challenges

The progress of the PVT technique, that we have briefly summarized, is remarkable indicating that it is becoming a very promising and simple method for particle manipulation and massive trapping. However, there are a number of challenges to enhance even more its technological potential. Main challenges, that in principle should be achievable, are: *a)* to enhance the pattern resolution to the nanometer scale, *b)* to increase the reproducibility and control of the deposition process, and *c)* to satisfactorily manipulate and deposit the particles from aqueous solutions favoring biotechnological applications. Further work in those directions should provide key advances in the near future.

Acknowledgments
This work was supported by MINECO, project number MAT2014-57704-C3

References

[1] Carrascosa M, García-Cabañes A, Jubera M, Ramiro J B, Agulló-López F 2015 *Appl. Phys. Rev.* **2**, 040605.
[2] Esseling M, Zaltron A, Sada C, Denz C 2013 *Appl. Phys. Lett.* **103**, 061115.
[3] Muñoz-Martínez J F, Elvira I, Jubera M, García-Cabañes A, Ramiro J B, Arregui C and Carrascosa M 2015 *Opt. Mater. Express* **5**, 1137.
[4] Jubera M, Elvira I, García-Cabañes A, Bella J L and Carrascosa M 2016 *Appl. Phys. Lett.* **108**, 023703.
[5] Arregui C, Ramiro J B, Alcázar A, Méndez A, et al. 2015 *J. Eur. Opt. Soc.* **10**, 15026.
[6] Elvira I, Muñoz-Martínez J F, Jubera M, García-Cabañes A, Bella J L, Haro-González P, Díaz-García M A, Agulló-López F and Carrascosa M, *Adv. Mat. Technol.* 2017,1700024
[7] Miccio L, Marchesano V, Mugnano M, Grilli S and Ferraro P 2016 *Opt. and Lasers in Eng.* **76**, 34.
[8] Muñoz-Martínez J F, M Jubera, J Matarrubia, A García-Cabañes, Agulló-López F and Carrascosa M 2016 *Opt. Lett.* **41**, 432.
[9] Esseling M, Zaltron, Horn W and Denz C 2015 *Laser and Photonics Rev.* **9,** 98
[10] Chen L, Fan B, Yan W, Li S, Shi L and Chen H 2016 *Opt. Lett* **41**,4558
[11] Chen L, Li S, Fan B, Yan W, Wang D, Shi L, Chen H, Ban D, Sun S 2016 *Scientific Reports* **6**, 29166
[12] Gazzetto M., Nava G., Zaltron A, Cristiani I, Sada C and Minzioni P, 2016 *Crystals* ***6***, 123

IOP Conf. Series: Journal of Physics: Conf. Series **867** (2017) 012031 doi:10.1088/1742-6596/867/1/012031

Convolution Operation of Optical Information via Quantum Storage

Zhixiang Li, Jianji Liu, Hongming Fan, Guoquan Zhang*

The MOE Key Laboratory of Weak Light Nonlinear Photonics, School of Physics and TEDA Applied Physics Institute, Nankai University, Tianjin 300457, China

*zhanggq@nankai.edu.cn

Abstract. We proposed a novel method to achieve optical convolution of two input images via quantum storage based on electromagnetically induced transparency (EIT) effect. By placing an EIT media in the confocal Fourier plane of the 4f-imaging system, the optical convolution of the two input images can be achieved in the image plane.

1. Introduction

Optical convolution operation is one of the most important technique in optical information processing. Theoretically, convolution of two input images in space domain can be achieved through multiplication of their Fourier spectra. In general, optical convolution can be achieved experimentally in a 4f-imaging system by multiplying the Fourier spectrum of an input image with that of another input image, which requires a specially designed filter placed in the confocal Fourier plane of the 4f-imaging system [1-2]. Here, we demonstrated a direct method to achieve optical convolution of two input images based on EIT effect.

2. Convolution Operation via EIT-based light pulse storage and retrieval

2.1 Theoretical analysis

Electromagnetically induced transparency effect is a result of destructive quantum interference among multiple atomic transition paths, it provides a powerful way for coherent manipulation of the interacting light fields and the optical properties of the atomic media. In a coherent atomic ensemble driven by a strong coupling beam, the absorption of a probe beam, even resonantly applied on an atomic transition, can be reduced or even eliminated under the EIT condition. According to the Kramers-Kronig relationship, the EIT effect will induce a strong spectral dispersion near the two-photon resonant condition, which was used to slowdown and stop light pulses propagating in the EIT media [3-5]. Storage and manipulation of light pulses carrying optical information [6-12] based on the EIT effect have attracted much interests in recent years.

By placing an EIT media in the confocal Fourier plane of the 4f-imaging system, one can firstly store the Fourier spectrum of the signal light pulse $E_S(x)$ carrying one of the input image in the EIT media by adiabatically switching off the Gaussian coupling beam $E_C(x)$, then the stored Fourier spectrum is read out by the Fourier spectrum of a readout beam $E_R(x)$ encoded with another input image. In this way, the retrieved light field $E_{S'}(x)$ is actually the multiplication of the Fourier spectra of the two input images:

$$E_{S'}(x) \propto E_S(x)E_R(x) \quad (1)$$

Thus the output in the imaging plane of the 4f-imaging system is then proportional to the optical convolution of the two input images

$$E_{S'}(\mathbf{r}_i) \propto E_S(\mathbf{r}_i) \otimes E_R(\mathbf{r}_i) \quad (2)$$

Note that we have not accounted for the finite numerical aperture of the 4f imaging system in the calculations.

2.2 Experimental verification of convolution operation

The proposed method was experimentally confirmed in a modified 4f-imaging system with a Pr^{3+}:Y_2SiO_5 crystal as the EIT media and two double-slits as input images. The slit width was set to be a=200 μm and the distance between two slits was d=1000 μm, respectively.

The experimental results are shown in figure 1, where (a) and (b) are the transverse intensity distribution of the signal and the readout beam in the image plane, while (c) is the transverse intensity distribution of the retrieved signal S'. Figure 1(c) with three peaks is a typical convolution operation of the information carried by the signal and the readout beams.

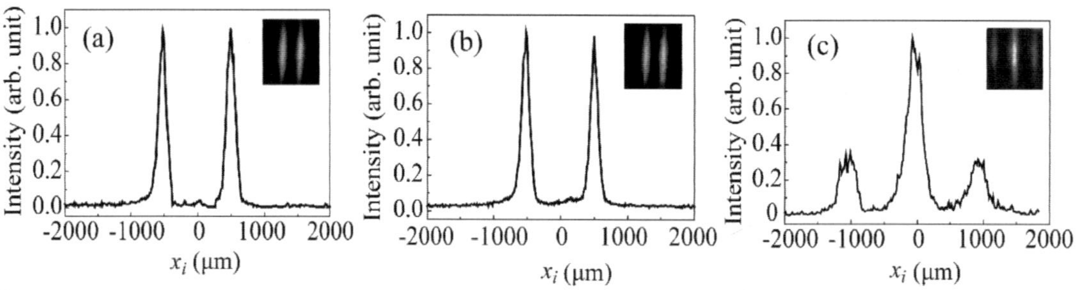

Figure 1. (a)-(c) are the transverse intensity distribution of the signal beam, the readout beam and the retrieved beam in the image plane, respectively. The insets show the corresponding intensity distribution recorded by a CCD.

References

[1] Lv N 2008 *Fourier optics* (Beijing: Mechanical Industry Press) 12-16
[2] Hecht E 2002 *Optics* (San Francisco: Addison Wesley) 532-542
[3] Fleischhauer M and Lukin M D 2000 *Phys. Rev. Lett.* **84** 5094–5097
[4] Hau L V, Harris S E , Dutton Z and Behroozi C H 1999 *Nature* **397** 594–598
[5] Turukhin A V, Sudarshanam V S, Shahriar M S, Musser J A, Ham B S, and Hemmer P R 2001 *Phys. Rev. Lett.* **88** 023602
[6] Liu C, Dutton Z, Behroozi C H and Hau L V 2001 *Nature* **409**, 490–493
[7] Pugatch R, Shuker M, Firstenberg O, Ron A and Davidson N 2007 *Phys. Rev. Lett.* **98** 203601
[8] Cho Y -W and Kim Y -H 2010 *Phys. Rev. A* **82** 033830
[9] Zhai Z, Li Z, Xu J and Zhang G 2013 *Phys. Rev. A* **88**, 035807
[10] Li Z, Liu J, Fan H, Liu J and Zhang G *Sci. Rep.* **7** 2361
[11] Vudyasetu P K, Camacho R M and Howell J C 2008 *Phys. Rev. Lett.* **100** 123903
[12] Shuker M, Firstenberg O, Pugatch R, Ron A and Davidson N 2008 *Phys. Rev. Lett.* **100** 223601

IOP Conf. Series: Journal of Physics: Conf. Series **867** (2017) 012032 doi:10.1088/1742-6596/867/1/012032

Nanoparticle Gratings for Compact Spectrometers: an Application of Photovoltaic Tweezers

JF Muñoz-Martínez[1, 2, 3], G Martín[3], A Alcázar[1], M Carrascosa[2]

[1]Universidad Politécnica de Madrid, Dpto. de Mecánica de Fluidos y Propulsión Aeroespacial, ETSIAE, 28040, Madrid, Spain

[2]Universidad Autónoma de Madrid, Dpto. de Física de Materiales, Facultad de Ciencias, 28049, Cantoblanco (Madrid), Spain

[3]Université Grenoble Alpes/CNRS, IPAG, F-38000 Grenoble, France

jf.munoz@upm.es

Abstract. Compact spectrometers are useful in fiber-based setups, portable devices or systems where weight and space are limited. On the other side, Photovoltaic Tweezers (PVT) are a simple and flexible optoelectronic technique for patterning and trapping nano- and microobjects in accordance with arbitrary light profiles. In this work, Al nanoparticle (NP) gratings assembled by PVT have been used for the fabrication of SWIFTS-based and dispersive spectrometers. A method for transferring the NP structures has been successfully adapted to deposit these gratings on top of Ti-diffused LiNbO3 wavguides (WG) without a significant loss of quality; the successful transference of NP patterns to arbitrary substrates remarkably extends the potential applications of PVT.

1. Introduction

Integrated optical spectrometers are an excellent option for reducing weight and power consumption in interferometric devices. Additionally, the almost total absence of moving parts provides high robustness and improved stability [1]. In fact, since integrated systems use to be intrinsically compact, they are a perfect option for applications in which weight and space are limited and/or the system undergoes severe vibrations or forces. These applications include portable devices or aerial spectroscopy systems set on unmanned aerial vehicles (UAV) and satellites [2].

Compact spectrometers are mainly classified in two groups: Fourier transform-based (or SWIFTS) and dispersive ones. The former family encompass those devices in which an interferogram is generated inside a waveguide (WG) and then it is sampled by scattering points placed on the WG surface. The spectrum is recovered using a Fourier Transform algorithm. The second group also confines the input light in a WG but they use a surface diffraction grating and relay optics to directly obtain the signal spectrum on a detector [1].

In this work, nanoparticle (NP) gratings assembled by Photovoltaic Tweezers (PVT) have been used for the fabrication of both kind of compact spectrometers.PVT are an emergent, flexible optoelectronic technique for patterning and manipulation of nano- and micro-objects in accordance with arbitrary light profiles [3]. It takes advantage of the evanescent electric fields that appear in photorefractive crystals with high bulk photovoltaic effect, like $LiNbO_3$:Fe, when they are suitably illuminated. PVT have proven to be a useful tool for controlled arrangement of bio-objects, making diffractive optical elements or producing plasmonic platforms.

2. Devices fabrication

In order to fabricate these devices, a series of Al NP linear gratings with different periods and fill factors were patterned via PVT. This was a two-step process. First, X- and Z-cut $LiNbO_3$:Fe crystals were illuminated with a two Gaussian beam interference in order to obtain the appropriate space-charge distribution. Afterwards, the crystals were introduced in a hexane suspension with the Al NP, which settled on the crystals surface making up fringes. Depending on the spectrometer configuration, NP strips were designed to be sparse and so, behave just as scattering centers (SWIFTS), or as fully dense linear diffraction gratings (dispersive).

These NP patterns were transferred on top of $LiNbO_3$ samples with several Y-junctions Ti-diffussed WG fabricated on its surface. This process was carried out following a method close to the one described in [4]. There was not a significant loss of quality between original and transferred patterns.

3. Spectrometers characterization

Regarding SWIFTS, the sample was attached to an optical bench consisting in a light detector and a Michelson interferometer. Two He-Ne laser beams, each coming from one arm of the interferometer, were injected in the Y-junction WG. The stationary wave was obtained by simple Fresnel reflection at the WG output. Scattered intensity of a single NP cluster was recorded as a function of the optical path delay, obtained by displacement of the mirror on one of the interferometer arms. We were thus able to detect the beams interference (see Figure 1a). However, we were unable to find the interferogram corresponding to a polychromatic light source due the combination of the wavepacket dispersion inside the WG, the reduced contrast of fringes in a SWIFTS configuration using simple Fresenel reflection and the large size of the scattering center.

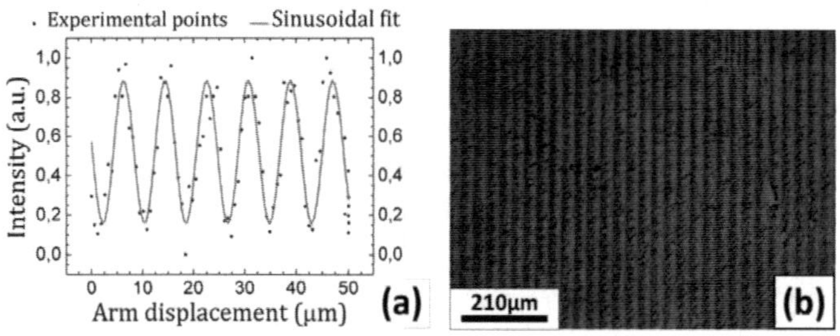

Figure 1. a) Scattering intensity of a single NP cluster as a function of the displacement of one of the Michelson arms. **b)** Al NP grating (period 30 μm) transferred on a $LiNbO_3$ WG.

For the characterization of the dispersive spectrometer, a 1.5 μm infrared laser source was employed. The key elements of the setup were a pair of mirrors, a cylindrical lens to collect the light extracted from the WG and a detector. The laser beam was divided using the mirrors and again injected in a Y-junction WG. The NP grating deposited on this sample was considerably thicker than the previous one (see Figure 1b) and it worked as a proper linear diffraction grating. Light was extracted from each arm of the Y-junction WG, collected by the lens and its interference recorded with the detector. Successful results were obtained, although the presence of different modes in the WG and defects in the NP grating affected negatively to the signal profile.

4. Future perspectives

This work present one of the first practical applications of PVT. Although the first results are not as good as the current state of the art, they are promising and expected to be improved by decreasing the size of the scattering centers. The fabrication of compact spectrometers by PVT has the advantage of being much faster than conventional methods such as focused ion beam (FIB) or Au-NP deposition. In addition, gratings made by PVT are easily reconfigurable, which means the WG can be reused. Finally, the successful transference of high-quality NP patterns to arbitrary substrates is a notable advance itself, as it remarkably broaden the potential applications of PVT.

Acknowledgments

This work was supported by MINECO, project number MAT2014-57704-C3. The first author was partially supported by a fellowship of the Universidad Politécnica de Madrid (RR01/2016).

References

[1] Martin G *et al. Proc. SPIE* 9516, Integrated Optics: Physics and Simulations II, 95160C (2015)
[2] Aerial Spectroscopy for Crop Monitoring [Internet] Ocean Optics 2014 Nov [cited 30 May 2016]. Available from: https://oceanoptics.com/aerial-spectroscopy-crop-monitoring/
[3] Carrascosa M *et al.* 2015 Applied Physics Review **2** 040605
[4] Li H *et al.* 2014 ACS Nano **8** (7), pp 6563-6570

IOP Conf. Series: Journal of Physics: Conf. Series **867** (2017) 012033 doi:10.1088/1742-6596/867/1/012033

EO properties and photorefractive two-beam coupling in AlN at 375 nm

Tomaž Kavaš, Aleksej Majkić, and Marko Zgonik

Faculty of Mathematics and Physics, University of Ljubljana, Ljubljana, Slovenia and J. Stefan Institute, Ljubljana, Slovenia

marko.zgonik@fmf.uni-lj.si

Abstract. Aluminium nitride (AlN) is a wide bandgap semiconductor with an intrinsic band-to-band absorption edge at ~200 nm wavelength. Therefore, the material is promising for optical use in the ultraviolet region. With a wurtzite-type, noncentrosymmetric crystal structure, AlN is also interesting for second order nonlinear optics and electro-optics. As grown crystals of AlN show relatively large absorption starting below 650 nm. With the material properties of AlN supporting the photorefractive effect we examined AlN as a photorefractive beam coupler. No effect was observed in the visible spectral range. Using the UV wavelength of 375 nm, where the absorption constant is (17 ± 2) cm^{-1} for the extraordinary polarization beam coupling was realized. The results show relatively weak and slow photorefractive effect.

1. AlN material properties

The nonlinear coefficients in AlN have been studied theoretically and experimentally in early AlN crystalline thin films. In recent years, significant progress has been made in the growth of good optical quality bulk crystals of AlN [1]. Due to impurities incorporation during growth the crystals are brownish in color and begin to absorb below 650 nm.

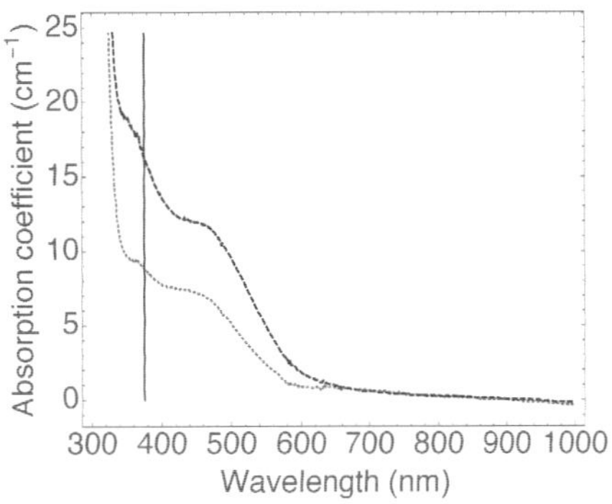

Figure 1. Absorption coefficient of bulk AlN crystal for extraordinary (dashed line) and ordinary polarization (dotted line). Measurements were performed on a spectrophotometer with a 1110 nm thick uncoated sample, with the refractive indices from Ref. [2]. Vertical line is drawn at 375 nm.

Large size samples allowed us to measure the optical nonlinear coefficient and two unclamped electro-optic tensor elements r_{33}^{T} = (1.16 ± 0.13) pm/V and r_{33}^{T} = (0.11 ± 0.02) pm/V at a wavelength of 633 nm. Miller theory can be used to estimate the photoelastic contribution to EO coefficients and their dispersion. The electro-optic coefficient $r_{ij}(\lambda)$ for the interacting optical beam at wavelength λ and the Miller-δ are related as

$$r_{ij}^{T,S} = \delta_{iij} \frac{2[\varepsilon_{ii}(\lambda)-1]^2[\varepsilon_{jj}^{T,S}(\infty)-1]}{\varepsilon_{ii}(\lambda)^2}, \tag{1}$$

With the use of refractive index data $\varepsilon_{ii}(\lambda)$ from Ref. [2], low frequency unclamped dielectric constants $\varepsilon_{33}^{T}(\infty)$ = 10.51 and $\varepsilon_{11}^{T}(\infty)$ = 8.44 from [3], and the clamped dielectric constants $\varepsilon_{33}^{S}(\infty)$ = 9.22 and $\varepsilon_{11}^{S}(\infty)$ = 7.84 that were determined by terahertz spectroscopy [4] we have calculated the dispersions shown in Fig. 2. As seen the photoelastic contribution $r^{T} - r^{S}$ in AlN is small.

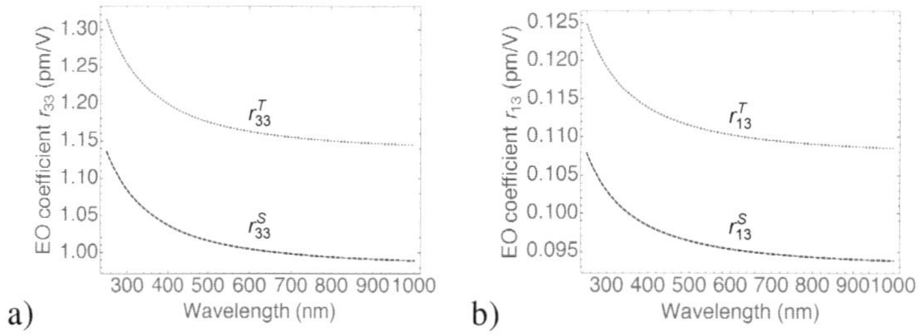

Figure 2. Dispersion of the EO coefficients of AlN crstal.

2. Two beam coupling

With the material properties of AlN supporting the photorefractive effect we examined AlN as a photorefractive beam coupler. Interference patterns with ***k***-vector parallel to the polar axis and diverse periodicities were created in the sample. Only extraordinary polarization was used in order to take advantage of the largest ***r*** coefficient. No coupling between the two beams was observed in the visible spectral range using the wavelengths of 633 nm, 532nm and 458 nm. By using the UV wavelength of 375 nm, however, where the absorption constant for the extraordinary polarization is 17 ± 2 cm^{-1}, beam coupling was observed. We set equal the intensities of the two impinging beams and measured the temporal dependence of the transmitted beams while chopping one of the beams with a repetition of 4 minutes. Because of the large absorption the single beam transmission was only about 10%. The coupling constants and response times are listed in Table 1. The photorefractive effect was found to be dominantly diffusion driven.

Table 1. Photorefractive two-beam coupling results at a wavelength of 375 nm.

Grating period (nm)	530	1000	1800
Coupling gain (cm^{1})	0.4 ± 0.05	0.5 ± 0.05	0.2 ± 0.1
Grating build-up time (s) at intensity 0.2 W/cm^2	6 ± 1	9 ± 1	5 ± 1
Grating build-up time (s) at intensity 0.04 W/cm^2	16 ± 2	18 ± 2	11 ± 2

References

[1] P. Lu, R. Collazo, R. F. Dalmau, G. Durkaya, N. Dietz, B. Raghothamachar, M. Dudley, and Z. Sitar, 2009 *J. Cryst. Growth* **312**, 58

[2] M. Rigler, J. Buh, M. P. Hoffmann, R. Kirste, M. Bobea, S. Mita, M. D. Gerhold, R. Collazo, Z. Sitar, and M. Zgonik, 2015 *Appl. Phys. Express* **8**, 042603

[3] T. Kim, J. Kim, R. Dalmau, R. Schlesser, E. Preble, and X. Jiang, 2015 *IEEE Trans. Ultrason. Ferroelectr. Freq. Control* **62**, 1880

[4] A. Majkić, A. Franke, R. Kirste, R. Schlesser, R. Collazo, Zlatko Sitar and M. Zgonik, 2017 *Phys. Status Solidi B*, DOI 10.1002/pssb.201700077

IOP Conf. Series: Journal of Physics: Conf. Series **867** (2017) 012034 doi:10.1088/1742-6596/867/1/012034

Fast novel nonlinear optical NLC system with local response

Andrey Iljin[1,*], Stefania Residori[2] and Umberto Bortolozzo[2]

[1]Institute of Physics, NAS of Ukraine, Prospect Nauki 46, Kyiv, Ukraine, 03028
[2]Institut de Physique de Nice, Université de Nice-Sophia Antipolis, CNRS, UMR7010, 1361 route des Lucioles, Valbonne, France, 06560

*the corresponding author: lgtc@iop.kiev.ua

Abstract. Nonlinear optical performance of a novel liquid crystalline (LC) cell has been studied in two-wave mixing experiments revealing high diffraction efficiency within extremely wide intensity range, fast recording times and spatial resolution. Photo-induced modulation of the LC order parameter resulting from *trans-cis* isomerisation of dye molecules causes consequent changes of refractive indices of the medium (Light-Induced Order Modification, LIOM-mechanism) and is proved to be the main mechanism of optical nonlinearity. The proposed arrangement of the electric-field-stabilised homeotropic alignment hinders the LC director reorientation, prevents appearance of surface effects and ensures the optical cell quality. The LIOM-type nonlinearity, characterised with the substantially local nonlinear optical response, could also be extended for the recording of arbitrary phase profiles as requested in several applications for light-beam manipulation, recording of dynamic volume holograms and photonic lattices.

1. Introduction

Elastic deformation in an LC medium appearing with the stimulated LC director reorientation spreads over large volumes and is limited actually with the cell boundaries only. During last decades a vast majority of nonlinear optical effects were observed in dye-doped LC mixtures, in which the dye molecules essentially enhanced the reorientation of the LC director [1]. Due to elastic forces, however, the nonlinear optical response appeared to be substantially nonlocal wiping off possibilities of many an application. On the contrary, the LIOM-type nonlinearity yields a strictly local response that has been shown in dye-doped chiral nematics LC (CLC) with short response times (less than 1 ms), high gain and spatial resolution of 1 μm [2]. Slow and stored light applications have been also demonstrated with this type of media [3]. The helical structure of the CLC, however, greatly affects the initial polarisation state of light waves. In this concern, the design of an NLC cell lacking orientational nonlinearities as well as being polarisation insensitive might have a lot of prospects.

2. Experiment and discussion

Cells were made of ITO-covered glass substrates, with homeotropically aligning polyimide coating with gaps of ~30 μm. The mixture 0.3 w.% of Methyl Red dye in 5CB nematic LC was used. An ac electric field (104 V @ 10 kHz) was applied to the cell in order to stabilise the homeotropic orientation of the LC director and hinder any light-stimulated reorientation of the LC director. All the results and observed effects were practically identical for different dye/LC combinations. Dynamic diffraction gratings were induced in LC cells with the ordinary two-wave mixing (pumping from Nd-

DPSSL laser, 532 nm, 1 ÷ 32 mW); the laser wavelength was close to the dye absorption maximum. The cell normal was approximately parallel to the bisector of the recording beams. Appearance of dynamic diffraction gratings was observed due to the self-diffraction as well as diffraction of the probe beam (He-Ne laser, 633 nm, 1 mW) (figure 1). Temporal development of the grating was measured via the first orders of the self-diffracting and the probe beams, with almost perfect reproducibility of the observed effects.

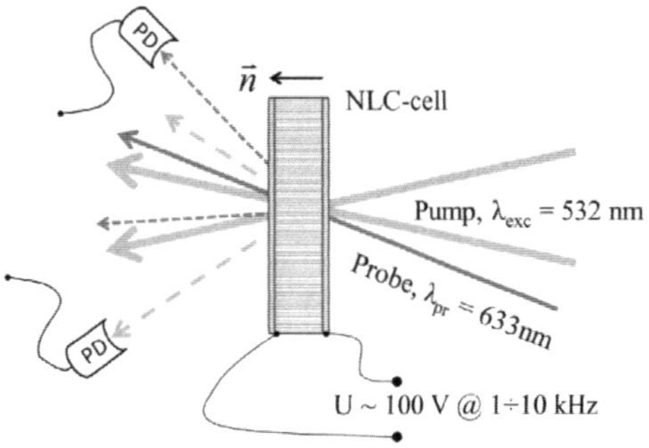

Figure 1. Experimental set-up. PD – photo-diode, $\vec{n}$ - LC director orientation.

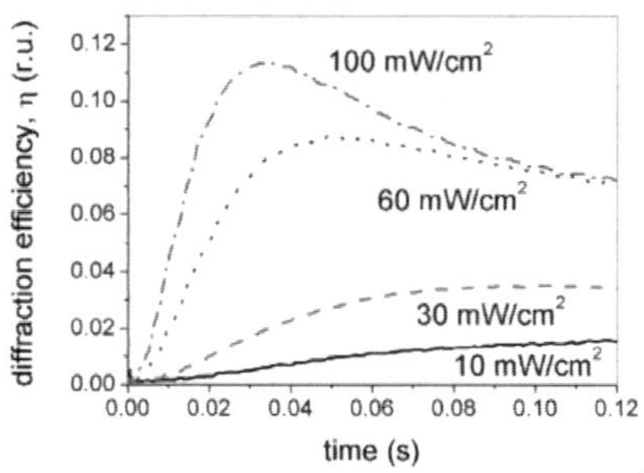

Figure 2. Typical traces of the 1st order self-diffraction signal for different pumping intensities.

The evaluation of the data shows that the intensity of the 1st order of self-diffraction passes through a maximum and, then, approaches saturation at some intermediate value (figure 2). The changes of the diffracted beam depend on the pump intensity – the higher the intensity, the faster the maximum is reached with the fastest recording times being of the order of 10 ms at least. Since MR-dye molecules are known for light-induced *trans-cis* isomerisation, such behaviour could be described in the frames of the LIOM-model [4]. Photo-induced *cis*-isomers disturb orientational LC ordering in their vicinity lowering the LC order parameter that, in turn, changes locally the refractive indices of the medium. In the current geometry the only ordinary refractive index is seen by the light wave and its modulation is determined as $\delta n_o = \alpha/2 c_{ph} \Delta n_0$, where Δn_0 is the initial LC birefringence, c_{ph} is the concentration of photo-induced dye molecules and α is the efficiency of the process. The consequent decrease of the diffracting beam intensity appears due to the saturation of the concentration of *cis*-isomers because of their quite long relaxation back to the *trans*-state.

3. Conclusions

Electric field stabilisation in the proposed LC-cell arrangement hinders the LC director reorientation and prevents appearance of surface effects. The LIOM-type nonlinearity provides substantially local nonlinear optical response, high efficiency and recording times well shorter than that of orientational nonlinearities. With the predicted high resolution and fast operation the designed NLC system is very promising for realization of various nonlinear optical phenomena such as the recording of dynamic volume holograms and photonic lattices.

References

[1] Khoo I C 2009 *Physics Reports* **471** 221
[2] Wei D, Iljin A, Cai Z, Residori S and Bortolozzo U 2012 *Optics Letters* **37** 734
[3] Wei D, Bortolozzo U, Huignard J P, Residori S 2012 *Optics Express* **21** 19544
[4] Iljin A 2011 *Mol.Cryst.Liq.Cryst.* **543** 143

IOP Conf. Series: Journal of Physics: Conf. Series **867** (2017) 012035 doi:10.1088/1742-6596/867/1/012035

Photorefractive Lithium Niobate crystals: light polarisation rotation highlighted by transmission Raman spectroscopy

N Kokanyan[1,2], M Bazzan[3], D Chapron[1,2] and M D Fontana[1,2]

[1]CentraleSupelec, Laboratoire Matériaux Optiques, Photonique et Systèmes, 2 rue E. Belin, 57070 Metz, France
[2]Université de Lorraine, Laboratoire Matériaux Optiques, Photonique et Systèmes, 2 rue E. Belin, 57070 Metz, France
[3]Physics and Astronomy Department, Universita di Padova, Via Marzolo 8, 35131 Padova, Italy

E-mail: ninel.kokanyan@centralesupelec.fr

Abstract. Transmission Raman spectroscopy experiments as a function on time were carried out on iron doped photorefractive congruent lithium niobate crystals. A big change with time in Raman spectra is observed when the incident light is polarized along ordinary axis. The mentioned breaking of Raman selection rules is originated by photorefractive properties of the crystal.

1. Introduction
Lithium Niobate (LN) is a multifunctional crystal with piezoelectric, ferroelectric and nonlinear optical properties [1]. The photorefractive (PR) effect is one of the most widely investigating properties in LN [2]. The main interest concerns application to holographic memory [3]. PR effect is largely enhanced in Fe-doped LN, [1-2] and is much reduced as the dopant is Zr 2% [4]. The main drawback of PR is the defocusing of the light beam inside the crystal [5]. Raman spectroscopy is a well-known technique providing information about vibrational and structural properties of a material. Here we use Raman spectroscopy to evidence the change of polarisation accompanying the light defocusing.

2. Experimental results and discussion
Transmission Raman spectroscopy measurements were performed on 0.03mol% and 0.05mol% iron doped LN crystals and spectra were recorded as a function of time. A diode laser with a radiation at the wavelength of 532nm was used. For X(YZ)X configuration the evolution of transmission Raman spectra with time is plotted in figure 1. For this configuration, only E[TO] modes are expected according to Raman selection rules. We observe that A_1[TO] lines unexpected by Raman selection rules and symmetry [6], are activated with time, and notably differ from data recorded in back-scattering [7]. The transmitted beam pattern with the same X(YZ)X configuration was recorded at different moments and is presented below (figure 2). The defocusing of transmitted beam with time is due to the polarization transmission from ordinary to extraordinary axis [8].

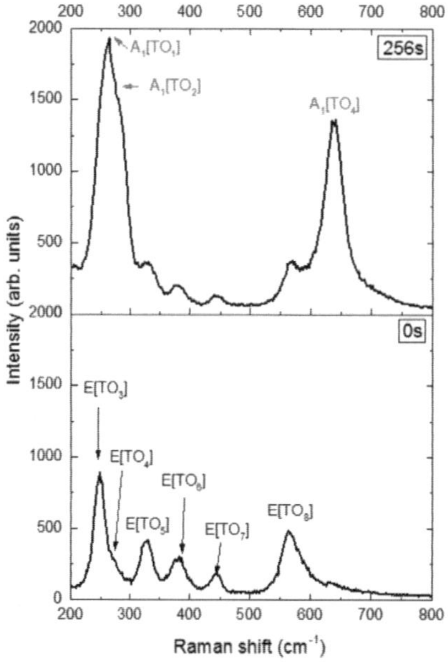

Figure 1. Transmission Raman spectra as a function of time in X(YZ)X configuration.

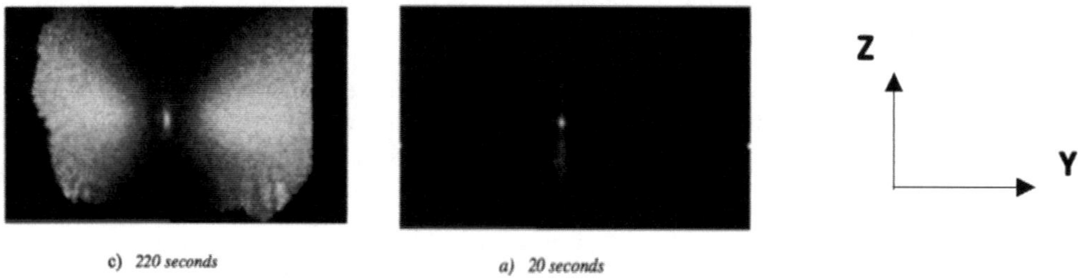

Figure 2. Recorded beam pattern of transmitted beam at 20 seconds and 220 seconds for X(YZ)X configuration

3. Conclusion

Transmission Raman spectra recorded with X(YZ)X configuration leads to breaking of Raman selection rules with activation of unexpected lines. This activation with time of forbidden A_1[TO] Raman lines for this configuration result from photorefractive internal field which causes the polarization conversion from ordinary to extraordinary polarization and is originated by photorefractivity of the crystal.

References

[1] Volk T R and Woehlecke M 2008, *Springer Series in Materials Science*.
[2] Günter and Huignard 1988, *JP Edts*. **1**, 131.
[3] Adibi A, Buse K, and Psaltis D 1999, *Opt Lett*. **24**, 652.
[4] N.Kokanyan, D.Chapron, E.Kokanyan, M.D.Fontana 2017, *J. of Appl. Phys*. **121**, 095103.
[5] M. Nippus and R. Claus, Z. Naturforsch 1978, Teil A **33**, 924.
[6] M.D.Fontana and P.Bourson 2015, Appl. Phys. Rev. **2**, 040602.
[7] M.Bazzan, N.Kokanyan, A.Zaltron, N.Argiolas, D.Chapron, M.D.Fontana 2013, *J. of Appl. Phys*. **114,** 163506.
[8] D W Wilson, E N Glytsis, N F Hartman and T K Gaylord 1992, *J. Opt. Soc. Am. B*, **9**, 1714

IOP Conf. Series: Journal of Physics: Conf. Series **867** (2017) 012036 doi:10.1088/1742-6596/867/1/012036

Photogalvanic self-pulsing of electro-wetting of LC droplet on LN-crystal

N Kukhtarev[1], T Kukhtareva[1], J Geng[2], XZ Zhang[2] and JJ Xu[2]
[1]Phys.Dep. Alabama A&M University
[2] The MOE Key Laboratory of Weak-Light Nonlinear Photonics, TEDA Institute of Applied Physics and School of Physics, Nankai University, Tianjin 300457, P. R. China

E-mail: knickoly@yahoo.com, zxz@nankai.edu.cn

Abstract. Photogalvanic current photoinduced by CW laser illumination of Fe-doped LN reveals quasi-periodic pulsing due to microplasmas discharges. This transformation of constant external influence into periodic response is interesting as example of self-excited systems, resembling pulsations of biological systems (including heart beats). Microplasmas is also interesting for potential applications in various fields, such as surface treatment, sterilization, water splitting, light sources, micro jets. Self-pulsing regime of microplasmas discharges in microhollow cathode discharges was observed and modeled in plasmas physics. In this paper, we analyze self-pulsing of photogalvanic current visualized by dynamic electrowetting in a simple LC-cell, formed by a nematic LC droplet placed on Fe-doped LN crystal. We have developed model of self-pulsing regime based on nonlinear discharge resistance. Experimental phase-space plot may be used to find parameters of the nonlinear discharge resistance. In the dynamic modeling contributions from both photogalvanic and pyroelectric effect are included. The analysis of dynamic regime reveals that a current pulse (in microsecond range) have asymmetric shape with extremely sharp rise with longer decay time. These self-pulsations are visualized by reflection of laser light from a LC droplet placed on LN-crystal. Reflected interference patterned was modulated by the dynamic electro-wetting effect.

1. Introduction

In description of self-pulsation we will start from the standard photorefractive equations that include pyroelectric contribution [1-4]. For modeling of nonlinear resistance will be used similar approach, that was developed in [5-6]. In this paper, we will expand the theoretical analysis to the high light intensities, where quadratic recombination is pronounced and the photogalvanic coefficient begins depend on the concentration of photo-excited charge carriers. This extension allows us to analyze the maximal photogalvanic voltage and current in practically important experiments when electrical discharges are pronounced.

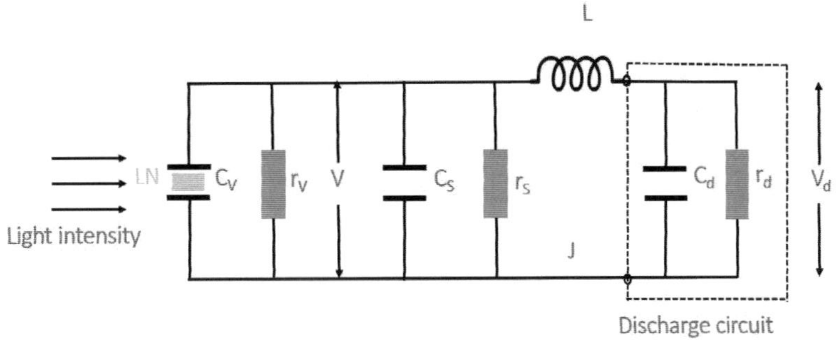

Figure1. Equivalent electric scheme of photogalvanic/pyroelectric crystal: C_v, r_v –capacitance and resistance of the crystal volume. C_s, r_s – effective capacitance and resistance of crystal surface. L – inductance of the effective discharge circuit. V, J, V_d, C_d, r_d - output voltage and current, voltage, capacitance and resistance of the discharge circuit.

IOP Conf. Series: Journal of Physics: Conf. Series **867** (2017) 012036 doi:10.1088/1742-6596/867/1/012036

We will consider the equivalent electrical scheme of photogalvanic/pyroelectric crystal (Figure 1) and include both the photogalvanic and the pyroelectric contributions to the crystal charging.

We will consider a semiconductor ferroelectric, with donor impurity with concentration N, which may be excited by laser illumination, producing electrons with concentration $\boldsymbol{n}$ in the conduction band and generating photogalvanic current along the crystal polar axis and with acceptors with concentration $\boldsymbol{N_A}$. In quasi-neutrality approximation and neglecting diffusion, equations for photoinduced voltage $\boldsymbol{V}$, current $\boldsymbol{J}$, photoinduced temperature change $\boldsymbol{\delta T}$ and concentration of electrons $\boldsymbol{n}$ can be written as:

$$\begin{aligned}
dV/dt &= -\frac{V}{(C_v+C_S)}\left(\frac{1}{r_v}+\frac{1}{r_s}\right) - AGIs(N-N_A-n)/(C_v+C_s) - d\delta T/dt\frac{pA}{(C_v+C_s)} + J/(C_v+C_s) \\
dn/dt &= s'I(N-N_A)-(sI+\gamma N_A)n-\gamma n^2 \\
dJ/dt &= (V_d-V)/L \\
dV_d/dt &= -\frac{V_d}{r_d C_d} - J/C_d \\
d\delta T/dt &+ \delta T/\tau_T = \alpha_T I
\end{aligned} \qquad (1)$$

Here, $\boldsymbol{C_s}$, $\boldsymbol{r_s}$ are effective capacitance and resistance of crystal surface, $\boldsymbol{C_d}$ is the capacity of the discharge circuit and $\boldsymbol{V_d}$ is the voltage on this capacitor, $\boldsymbol{L}$ is an effective inductance of the discharge circuit, $\boldsymbol{r_d}$ is effective discharge resistance, $\boldsymbol{\delta T}$ is photoinduced change of temperature, $\boldsymbol{p}$ is the pyroelectric coefficient, $\boldsymbol{\tau_T}$ is thermal relaxation time, $\boldsymbol{\varepsilon}$ is the effective dielectric constant and $\boldsymbol{s'=sF/\omega}$, $\boldsymbol{F}$ is the quantum efficiency, and $\boldsymbol{\omega}$ is photon energy, $\boldsymbol{\alpha_T=(\alpha+\alpha h)/(\rho\ C_h)}$ is the effective heat absorption coefficient, $\boldsymbol{\alpha = s(N-N_A-n)}$ is an optical absorption coefficient, $\boldsymbol{\alpha_T}$ is thermal absorption constant due to IR absorption, $\boldsymbol{\rho}$ is mass density, $\boldsymbol{C_h}$ is heat capacity.

Here $\boldsymbol{N}$, $\boldsymbol{N_A}$ and $\boldsymbol{n}$ are concentrations of donors, acceptors and electrons, respectively, $\boldsymbol{\mu}$ is mobility of electrons, $\boldsymbol{G}$ is photogalvanic coefficient, and $\boldsymbol{J}$ is the total current density, $\boldsymbol{s}$ is the absorption cross-section of photons, $\boldsymbol{s'=sF/\omega}$, γ is recombination coefficient, and $\boldsymbol{I}$ is intensity of light.

Denoting $\boldsymbol{A}$ and $\boldsymbol{d}$ as the electrode area and crystal thickness, the internal (volume) resistance of the illuminated crystal can be written as $\boldsymbol{r_V = d/(\sigma A)}$, where $\sigma = e\mu n$ is the electrical conductivity. For description of high-intensity case, we consider photogalvanic coefficient $\boldsymbol{G'}$ depends on concentration of photoexcited electron concentration $\boldsymbol{n}$:

$$\boldsymbol{G' = G\,(1 + k\,n)} \qquad \textbf{(2)}$$

where $\boldsymbol{k}$ is nonlinear photogalvanic coefficient.

To analyze regime of self-pulsation, we will consider steady-state for generation of photoelectrons $\boldsymbol{n}$. Put $\boldsymbol{C_d=0}$ and ignore (for simplicity) pyroelectric contribution. In this case equations will be reduced to (taking into account, that now $\frac{V_d}{r_d C_d} = -J$):

$$\frac{dV}{dt} = -\frac{V}{\tau} - AG' I\alpha / C + J / C$$
$$\frac{dJ}{dt} = -(J r_d + V) / L \tag{3}$$

here α is absorption coefficient, $C = C_s + C_v$.

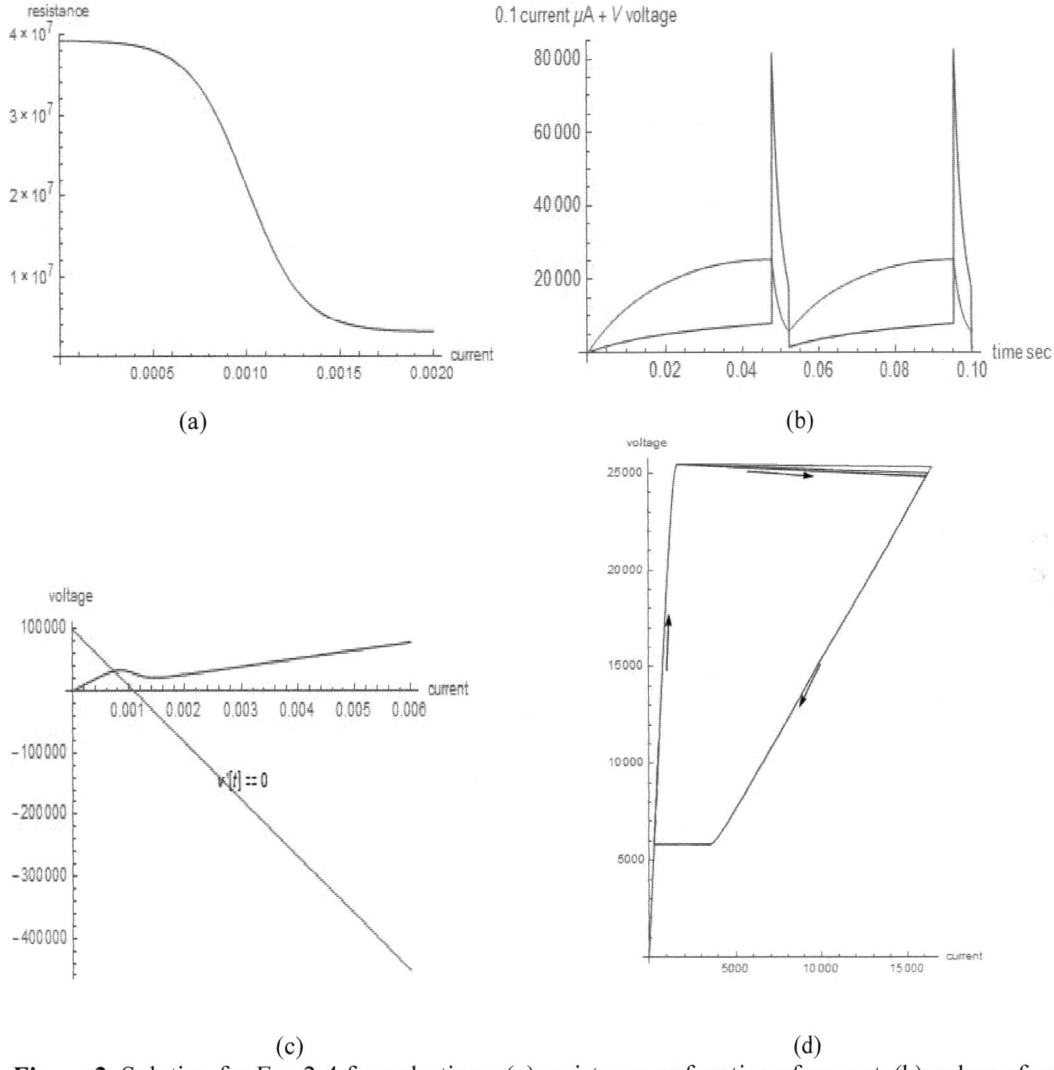

Figure 2. Solution for Eqs.2-4 for pulsations: (a) resistance as function of current, (b) pulses of current and voltage, (c)static V-A characteristic, (d) dynamic V/A characteristic. Period of pulsation T = 0.05 s. (Experiment on Sun T = 0.04 s)

IOP Conf. Series: Journal of Physics: Conf. Series **867** (2017) 012036 doi:10.1088/1742-6596/867/1/012036

To describe self-pulsations we will chose nonlinear bulk resistance r_d on current J as:

$$r = (a - b)\, tanh((J - i)/p) \qquad (4)$$

Choosing parameters as: **a+b=(2.12+1.812) 10^7 Ω, i=10^{-3} A, p=3*10^{-4} A , τ = 3.08*10^{-5} s, C = 7*10^{-10} F,** ***AG'I α /C=2.1*10^7 V/s*** , we can get solution of Eqs.2-4, Figure 2 is what we got using Mathematica software.

2. Experimental results

Simple experimental setup for detection of electrical and optical self-pulsation is shown in Figure 3.

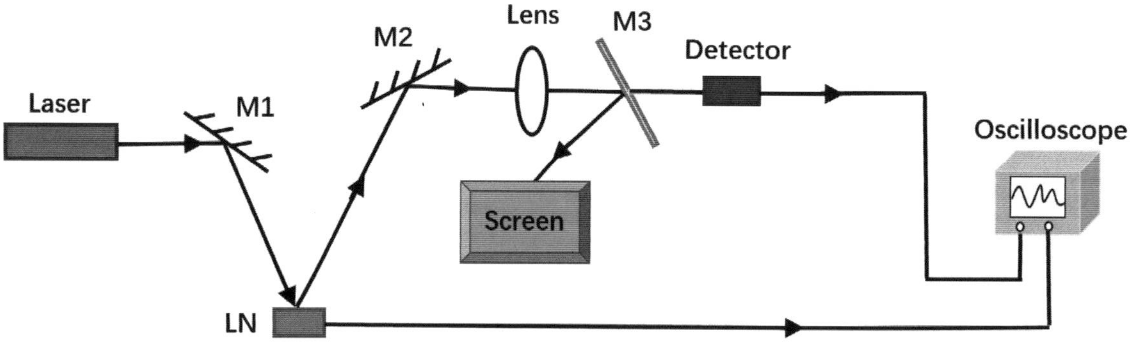

Figure 3. Experimental setup for observation of electrical and optical pulsation

We have used blue CW laser and Fe-doped C-cut LN crystal. Nematic Liquid crystal droplet (~ 5 mm diameter) was placed by pipette on the C-surface of crystal. Reflected beam was projected on the screen, with part of the beam registered by detector. Another channel of oscilloscope was connected to the electrode (one is gridded), covering C-surface. One of reflected beam from LC droplet dancing images is shown in Figure 4.

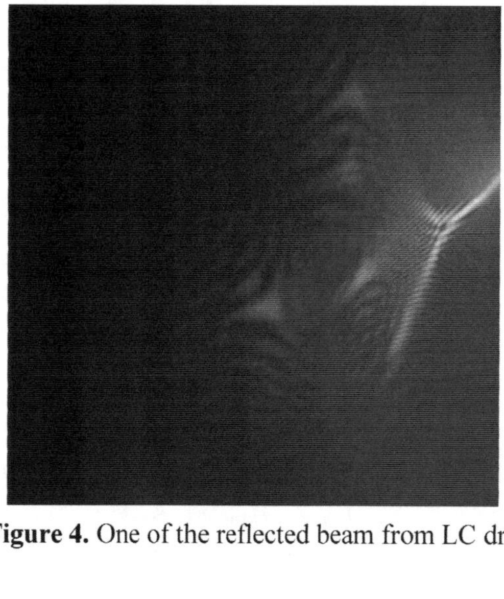

Figure 4. One of the reflected beam from LC droplet dancing images.

IOP Conf. Series: Journal of Physics: Conf. Series **867** (2017) 012036 doi:10.1088/1742-6596/867/1/012036

Examples of electrical pulses from the Sun-illuminated LN crystal are shown in Figure 5 (single pulse with 30 A amplitude) and Figure 6 (self-pulsation under Sun illumination).

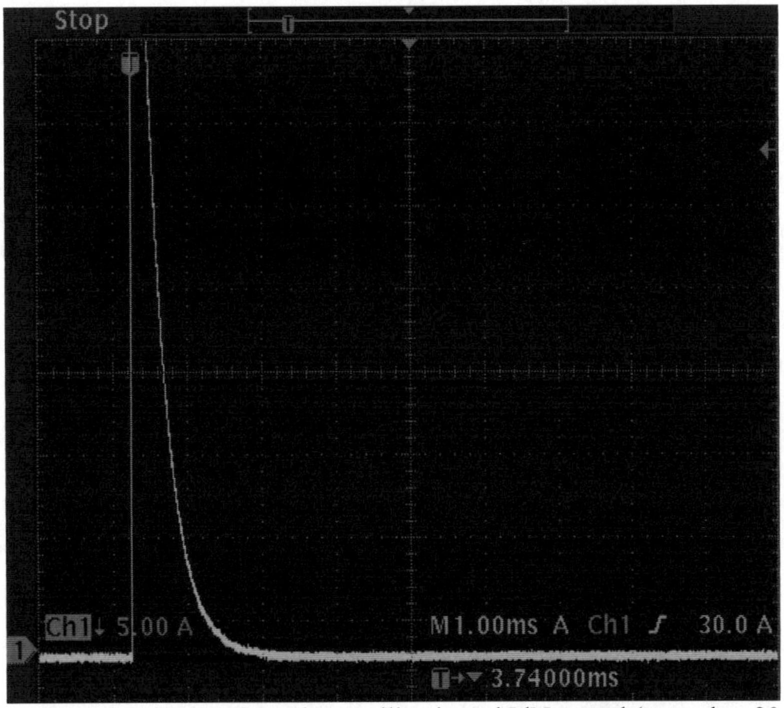

Figure 5. Current pulse from the sun-illuminated LIN crystal (more than 30 A)

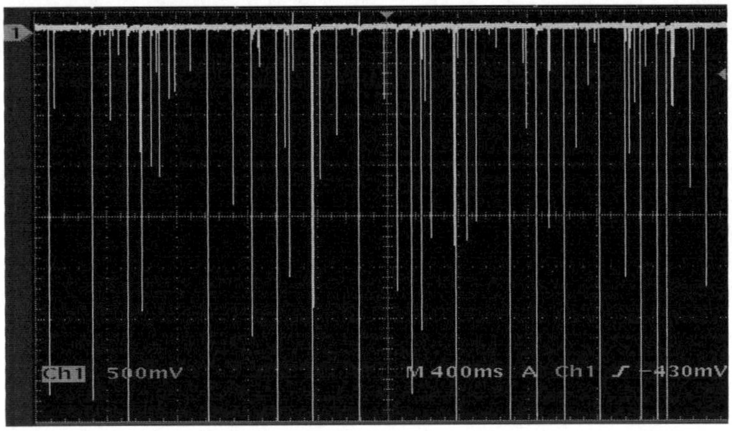

Figure 6. Electrical pulsations picked-up by grid electrode from 0.1 % Fe:LN illuminated by the concentrated solar light.

3. Discussions and conclusion

We have observed self-pulsations of electrical current, photo-induced in Fe-doped LN crystal monitored by electro-wetting effect from the drop of the nematic liquid crystal placed on C-surface of LN.

Theoretical model of the electrical self-pulsation is based on photogalvanic/pyroelectric charging of LN crystal that have nonlinear (N-type) volt-ampere characteristic (similar to discussed in [5]). Numerical simulations are illustrated by the phase-plot diagrams.

Optical pulsations of the laser beam reflected from the LC droplet coincides with the electrical pulsations. Dancing Interference patterns can be explained as single-beam parametric dynamic holograms, as parameters of electro-wetting are modulated by the same laser beam. Period of pulsations, induced by concentrated Sun illumination (~ 0.05 s) fit well with theoretical estimations (~0.04 s).

Illumination by laser and Sun light generates CW photogalvanic current with contribution from pyroelectric effect during transient heating when temperature changes in time. These currents create charge accumulations mainly on the C-surfaces until break-down voltages produce discharges.

The discharges generated by light may be mainly via the surface area with enhancements, due to plasma formation in the surrounding air. Possible contribution to the electrical discharges may come from the dynamic domain reversal [6].

Further investigation of electro-wetting pulsations may help not only for detection but also be used for characterization of electrical pulses.

References

[1] Krasnoholovets V, Kukhtarev N and Kukhtareva T 2006 *Int. J. Mod. Phys. B* **20** 2323
[2] Gnatenko Y P, Kukhtarev N, Faryna I O, Volkov V I, Bukivskij P M, Gamernyk R V, Rudenko V I and Paranchych S Y 2003 *J. Appl. Phys.* **94** 4896
[3] Beyer O, Breunig I, Kalkum F and Buse K 2006 *Appl. Phys. Lett.* **88** 051120
[4] Kukhtarev N, Kukhtareva T, Edwards M E, Jones J, Wang J, Lyuksyutob S F and Reagan M A 2005 *J. Appl. Phys.* **97** 054301
[5] Chabert P, Lazzaroni C and Rousseau A 2010 *J. Appl. Phys.* **108** 113307
[6] Yang S Y, Seidel J, Byrnes S J, Shafer P, Yang C H, Rossell M D, Yu P, Chu Y H, Scott J F, Ager J W, Martin L W and Ramesh R 2010 *Nature Nanotechnology* **5** 143

Fabrication of p-type lithium niobate crystals by molybdenum doping and polarization

Tian Tian[1,*], Yongfa Kong[2,3], Hongde Liu[2], Shiguo Liu[2], Wei Li[3], Shaolin Chen[3], Jiayue Xu[1]

1. Institute of Crystal Growth, School of Materials Science and Engineering, Shanghai Institute of Technology, Shanghai 201418, China
2. School of Physics, Nankai University, Tianjin 300071, China
3. MOE Key Laboratory of Weak-Light Nonlinear Photonics and TEDA Applied Physics Institute, Nankai University, Tianjin 300457, China

Corresponding author's e-mail: tiant@sit.edu.cn

Abstract. The lack of p-type lithium niobate limits it serving as an active material. A series of Mo-doped and pure congruent lithium niobate crystals were grown by Czochralski method under different polarization conditions. Their dominant carrier species were characterized by holographic experiment. The results showed dominant charge carrier species may be changed from electrons to holes when lithium niobate crystal was doped with Mo ions and polarized under the current of 70mA for 30 minutes. It indicated that p-type lithium niobate crystal could be fabricated by Mo-doping and suitably controlling the polarization condition. Mo-doped lithium niobate crystals can be a promising candidate for active components.

1. Introduction

The goal of integrated optics is to assemble and connect many waveguide based components with different functionality on a single chip. For this, lithium niobate ($LiNbO_3$, LN) is a favorite base material and has been dubbed by many as the "silicon of photonics" [1, 2]. However, lithium niobate practically always serves as a passive material[3, 4]. A truly all-light photonic chip could be fabricated only if passive and active components can be integrated on one and the same substrate. A p-n junction is the basic unit for an active component. By now, one of the main obstacles is the lack of p-type lithium niobate.

As we know, the electrical charge carriers in nominally pure LN are electrons and holes[5], the amount of which can be changed through doping techniques and inducing various defects. But it is unfortunately that the currently used dopants are unsuitable to produce p-type conduction[6]. Though the dominant carrier in the highly iron-doped $LiNbO_3$ (LN: Fe) could be changed to hole by thermal-electric oxidization, it extremely damaged the crystal for the leaving of lithium ions[7]. Obviously, it is needed to search for suitable dopants or precise control of defects to obtain p-type $LiNbO_3$.According to our previous work[8], a novel hexavalent molybdenum doped $LiNbO_3$ crystals (LN: Mo) allow for excellent characteristics, which was attributed to Mo^{6+} ions prefer to occupy regular Nb-sites rather than Li-sites due to their high valence and produce new defects Mo_{Nb}^{+}. Besides, the photorefraction could be further enhanced when the polarization current is improved with the same polarization time.

It means the polarization had a great influence on the defects that serve as UV photorefractive centers, from which holes may be excited [9].

2. Experiments

The optimum doping concentration of MoO_3 is 0.5mol%. For this present investigation, a series of congruent $LiNbO_3$ crystals doped with 0.5mol% Mo and pure congruent $LiNbO_3$ crystals (CLN) were grown along the *c* axis with the conventional Czochralski method. The [Li]/[Nb] composition was chosen as 48.38/51.62. As normal polarization condition applied for our crystals was with 70mA for 15minutes, polarization conditions were adjusted with higher current or longer time, labeled as $LN:Mo(70,15)$, $LN:Mo(70,30)$, $LN:Mo(125,15)$, and $CLN(70,30)$, respectively. The former and latter number in the bracket represents the polarization current and time in minutes, respectively. After polarization and annealing treatment, 3.0 mm and 1.0 mm thick *y*-oriented plates were cut and polished to optical grade.

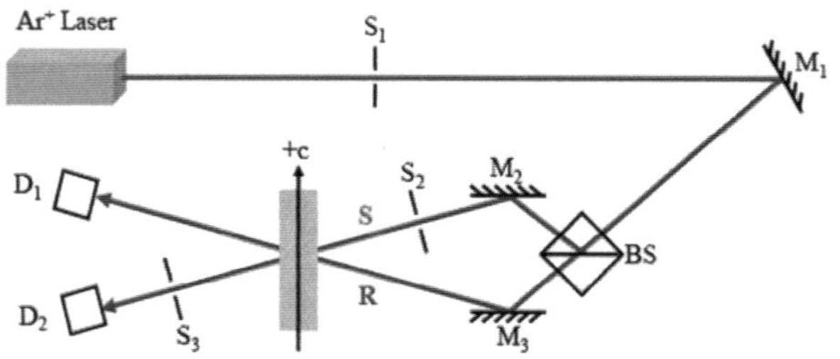

Fig. 1. Schematic diagram of the configuration for holographic experiment: M1–M3, mirrors; BS, beam splitter; S1–S3, shutters; D1–D2, detectors.

We judged the photo-induced carrier type of these samples by holographic experiment at 351nm, which was provided by an Ar^+ laser. The experimental setup is illustrated schematically in Fig. 1. The laser light was split into two equal beams (*e*-polarized, intensity per beam 320 mW/cm^2, diameter 1.5 mm), which would be reflected by two reflectors and formed signal beam (S beam) and reference beam (R beam). S and R beams located in the +c and –c direction of y-cut $LiNbO_3$ crystals, respectively. The dominant carrier type in $LiNbO_3$ crystals can be judged by comparing R curve with S curve. The judge principle are described in references[10]. In one word, if S curve decays slower than R curve, the dominant carriers are holes. On the contrary, electrons play the leading role.

3. Results and Discussions

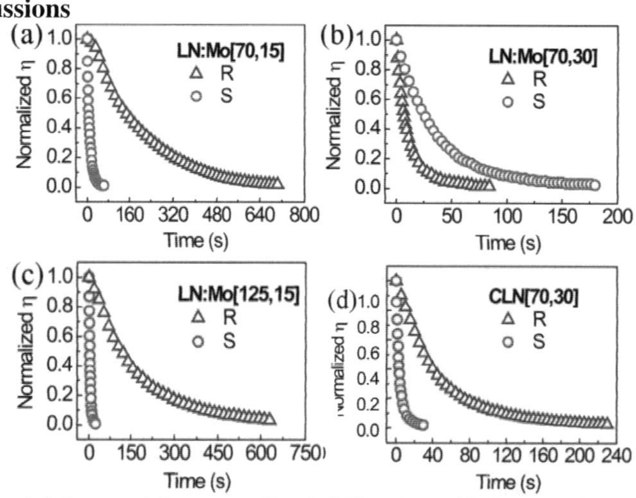

Fig. 2. Exponential decay of the normalized diffractive efficiency η for LN:Mo and CLN with different polarization conditions.

The exponential decay curves of η-t for LN: Mo polarized with various polarization conditions as shown in Figs. 2(a)-(d). It was evident that curve S was upper curve in Fig. 2(b), which proved that holes were the dominant carriers in LN : Mo(70,30). As presented in Figs. 2(a), (c), it was found that if 15 minutes was chosen as the polarization time, the increment of polarization current had no effect in altering the dominant carriers from electrons to holes. For comparisons, the type of dominant carriers in CLN crystals polarized under 70 mA for 30minutes were also examined. As shown in Fig. 2(d), the dominant carriers of both CLN crystals were electrons as curve R was upper curve, it indicated that the dominant carriers of CLN crystals could not be changed to holes by extending the polarization time to 30minutes with the current of 70 mA, which was suitable for LN: Mo crystals. Consequently, p-type LN crystals can be fabricated with the joint action of Mo-doping and suitable polarization proceeding. Moreover, utilizing this method one can grow a perfect bulk crystal without any damage as happened in thermal-electric oxidized highly doped LN: Fe crystals.

4. Conclusions

A series of LN: Mo crystals and CLN crystals were grown under different polarization conditions and their dominant carrier species were characterized by holography method. The dominant charge carrier species can be changed from electrons to holes when LN: Mo polarized under the current of 70 mA with extending the time to 30 minutes. But the increasing of polarization current and time has no effect in changing the dominant carriers to holes for LN: Mo and CLN crystals. Thus, p-type LN crystal could be fabricated by Mo-doping and suitably controlling the polarization condition while growing the crystal. LN: Mo can be a promising candidate for active components.

Acknowledgments

This work was supported by the National Natural Science Foundation of China (51572175 and 61605116), Jiangsu Planned Projects for Postdoctoral Research Funds (1501131C), Shanghai Planned Projects for Young Teacher Training of Universities (ZZyy15087).

References

[1] Kösters M, Sturman B, Werheit P, Haertle D and Buse K 2009 *Nature Photon.* **3** 510
[2] Broderick N *Nature Materials Update* 2002 Nov
[3] Sohler W, Hu H, Ricken R, Quiring V, Vannahme C, Herrmann H, Büchter D, Reza S, Grundkötter W, Orlov S, Suche H, Nouroozi R and Min Y 2008 *Opt. Photon. News* **19** 24
[4] Kienle F, Lin D, Alam S, Hung H, Gawith C, Major H, Richardson D and Shepherd D 2012 *J. Opt. Soc. Am. B* **29** 144
[5] Orlowski R and Kraetzig E 1978 *Solid State Commun.* **27** 1351
[6] Pei Z, Hu Q, Kong Y, Liu S, Chen S and Xu J, 2011 *AIP Adv.* **1** 032171
[7] Falk M, Woike T and Buse K, 2007 *J. Appl. Phys.* **102** 063529
[8] Tian T, Kong Y, Liu S, Li W, Wu L, Chen S and Xu J 2012 *Opt. Lett.* **37** 2679
[9] Liu F, Kong Y, Li W, Liu H, Liu S, Zhao Y, Liu H, Chen S and Xu J 2010 *Opt. Lett.* **35** 10
[10] Wang H, Wen J, Li J, Wang H and Jing J, 1990 *Appl. Phys. Lett.* **57** 344

Nanoparticle manipulation and trapping by the synergy between the photovoltaic and pyroelectric effects

J Faba[1], A Puerto[1], J F Muñoz-Martínez[1,2], A Méndez[2], A Alcazar[2], A García-Cabañes[1] and M Carrascosa[1]

[1]Dep. de Física de Materiales, Universidad Autónoma de Madrid, 28049 Madrid, Spain
[2]Dep. de Mecánica de Fluidos y Propulsión Aeroespacial, Universidad Politécnica de Madrid, 28040 Madrid, Spain

Abstract: We have investigated the combined action of the pyroelectric and photovoltaic effects exhibited by *z*-cut $LiNbO_3$:Fe substrates for particle trapping and patterning. The results suggest that a suitable combination of the PY and PV effects can improved some pattern properties such as particle density or pattern contrast.

m.carrascosa@uam.es

1. Introduction

Particle manipulation and trapping based in the electrical properties of ferroelectric crystals have been recently proposed and demonstrated. On the one hand, the pyroelectric (PY) effect generates patterned electric fields in $LiNbO_3$ domain structures [1] by thermal heating or cooling. On the other hand, the bulk photovoltaic (PV) effect generates light induced electric fields in accordance with the illumination patterns in certain doped ferroelectrics such as $LiNbO_3$:Fe (see [2] and references therein). In both cases, the induced fields allow particle manipulation via electrophoretic (EP) or dielectrophoretic (DEP) forces. However, the combined action of both effects have not been explored so far. In fact, in some configurations such as highly *z*-cut Fe-doped $LiNbO_3$ under high light intensities, both effects can appear simultaneously affecting particle trapping techniques. Furthermore, one can wonder about the possibilities of simultaneously using PY and PV effects to improve particle manipulation

In this work, we investigate the combined action of both effects presented by Fe doped $LiNbO_3$ substrates. To this end, we have developed a simple two-step strategy: *i)* illumination at low or moderate intensity generates a patterned electric field via the photovoltaic effect and, *ii)* subsequently, particle deposition from a cold suspension at a temperature several degrees lower than room temperature. The role of different parameters such as temperature variation (ΔT), or illumination time have been studied, and the results have been favorably compared with simple theoretical calculations. The experimental results suggest that a suitable combination of the PY and PV effects can improve some pattern properties such as particle density or pattern contrast. Other alternative experimental situations with the simultaneous action of both effects are briefly discussed.

2. Experimental methods

As ferroelectric substrates we have used 1 mm thick z-cut LiNbO3:Fe crystals with 0.09 wt% iron impurities. Nanoparticle (NP) patterns have been obtained in a two-step process [2]. First, the crystal sample is illuminated with a light pattern obtained using a spatial light modulator to induce PV electric fields. The light intensity is kept low enough to avoid any temperature change ($\Delta T < 1$ °C) and so, any PY effect. Next, without illumination, the crystal is introduced in a Ag NP suspension to fabricate the pattern. The average particle diameter is 100 nm. The suspension temperature is kept lower than room temperature to cool the crystal inducing PY effect during particle deposition on its surface.

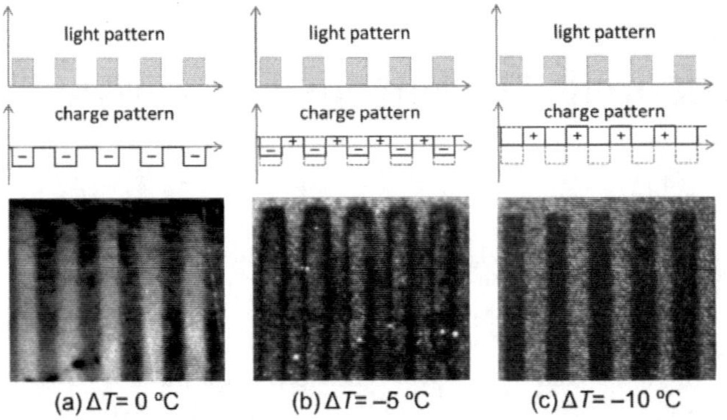

Figure 1: Photographs of Ag NP patterns deposited on the +*c* surface with a fringe light pattern at different temperature changes: (a) ΔT= 0 ºC, (b) ΔT= –5 ºC, (c) ΔT= –10 ºC. Schemes of the corresponding light and charge patterns are shown at the top of each photograph.

3. Results and discussion

In the applied strategy, we uncouple the PV and PY effects to find a configuration for which the combined action provide any advantage for particle trapping and patterning. As stated above, during illumination a patterned PV electric field is obtained. In the second step, a spatially homogeneous PY field is induced by cooling the sample. Hence, in the suspension, particles undergo the electric PV+PY fields and settle on the sample. The PV and PY (depolarization) surface charges have opposite sign, giving rise to a competition between both effects. To illustrate this competition we have developed a series of three experiments with the same illumination fringe pattern. The temperature decrease of the NP suspension with regard to the sample are ΔT =0 (for reference), –5 ºC and –10 ºC. The obtained NP patterns are shown in Figure 1a, 1b and 1c, respectively. In each figure schematics of the light pattern and the generated surface charge at the +*c* crystal face are also shown to discuss the physics. In figure 1a the typical reported PV-induced NP pattern for *z*-cut, replica of the light pattern, is obtained [3]. In Figure 1b, with ΔT = –5, we can see a double periodicity, i.e. additional particle fringes corresponding to the dark regions of the light pattern. This can be understood looking at the schemes: the magnitude of the surface charge densities generate by the PV (green dash line) and PY effects (red dash line) are similar (although with a different spatial distribution) and with opposite sign. So, the net charge (black solid line) alternates from negative to positive values. When using neutral particles, i.e. DEP forces act on them, they deposit on the substrate no matter the sign of the surface charge. Finally, in Figure 3c, with ΔT = –10, the PY effect is high enough to fully compensate the PV-induced charge and so, particle deposit only in the dark regions. Moreover, the pattern contrast is very good in this case, suggesting that this combination is useful to improve the pattern visibility. We have obtained similar trends for a set of experiments with a constant suspension temperature, ΔT = –5, but varying the illumination time and so, the magnitude of the PV effect with regard to the PY effect. We are now developing simulations using a model essentially based in the mechanisms just described. We expect to predict in detail the trapping behavior under PV+PY effects.

In fact, a number of interesting alternative situations can be explored and simulated. As an example, we have found that a focused Gaussian beam with light intensity $I \geq 10$ W/cm^2 produces in our samples a local temperature increase above 15 ºC, originating simultaneously PV and PY effects. In this case, both effects are local and generate charges with the same sign. NP trapping experiments to elucidate the main features of this situation, inherent to high light intensities and strong Fe-dopings, are now in progress.

Acknowledgment
This work was supported by MINECO, project number MAT2014-57704-C3

References

[1] Grilli S and Ferraro P 2008 *Appl. Phys. Lett.* **92**, 232902.
[2] Carrascosa M, García-Cabañes A et al. 2015 *Appl. Phys. Rev.* **2**, 040605.
[3] Muñoz-Martínez J F, Elvira I, Jubera M, García-Cabañes A, Ramiro J B, Arregui C and Carrascosa M 2015 *Opt. Mater. Express* **5**, 1137.

Spectral photosensitization of optical anisotropy in poly(vinyl cinnamate) solid films

V M Kozenkov[1], A A Spakhov[1], V V Belyaev[1,2], D N Chausov[1], O V Chausova[3], V G Chigrinov[4]

[1]Moscow Region State University, 10a Radio st., 105005 Moscow, Russia
[2] The Peoples' Friendship University of Russia (RUDN University), 6 Miklukho-Maklaya st., 117198, Moscow, Russia
[3]University of Technology, 42 Gagarina st., 141070, Korolev, Russia
[4]Hong Kong University of Science and Technology, Clear Water Bay Road, Kowloon, Hong Kong, China
E-mail: d.chausov@yandex.ru

Abstract. We observed possibilities and properties of sensitized photo induced optical anisotropy formation in amorphous poly (vynyl-cinnamate) films (PVCi) and its derivant poly (vynyl-4-methoxy-cinnamate) (PVMCi) under polarized light, including the one which is not absorbed by the supermolecules of the polymeric material. The effect of the induced optical anisotropy involves transferring the energy of electronic excitation from the donor molecules (Michler's ketone or 2-benzoyl-methylene-3, methyl-β-naphthosol) to the scavenger's molecules - PVCi or PVMCi) and photo-topochemical ring formation of cinnamate units in polymeric supermolecules. The discovered photo-induced anisotropy in solid PVCi and PVMCi films provides sensitized photo orientation of low-molecular thermotropic liquid crystals.

Key-words: photo-induced optical anisotropy, double refraction, poly (vynyl-cinnamate) photosensitization.

1. Introduction

The goal of current work is to analyze the formation of sensitized photoinduced optical anisotropy in amorphous poly(vynyl-cinnamate) films (PVCi) and in its derivant poly(vynyl-4-methoxy-cinnamate) (PVMCi) [1-3] under the influence of polarized light as well as to find physical mechanisms of this process so that it could improve the materials efficiency in different optics and optical information processing tasks [4-6].

2. Results and Discussion

The initial kinetics sections of POA formation within PVMC$_i$ layers, sensitized by BNT naphthotiazoline are shown in fig. 1 a–c for different wave-lengths when λ_{exp} = 436, 405 and 365 nm. The radiation with a wave-length of 436, 405 nm is absorbed only by BNT molecules and the latter wave-length lies within the maximum BNT absorption (~ 400 nm). As can be seen in fig. 1a and 1b, in PVMC$_i$ (curve 1) no POA is formed without a sensitizer. When activated with the wave-length of 365 nm, where the UV radiation is absorbed by both components, the formation speed has its maximum without any sensitizer (fig.1c, curve 1). In this case, if the sensitizer concentration increases, the POA formation speed decreases substantially (curve 2 and 5). This phenomenon has an inverse trend when the wave length is 436 nm, which is presented in substantial increase of photo-induced double refraction (DR) together with the BNT concentration increase in the PVMCi layer (fig.1a curve 2 and 5). For the radiation at 405 nm the effect of DR induction speed increase (fig.1b, curve 2 and 3) can be observed only with low concentration ~ 0.5 – 1.0 % and decreases when the concentration gradually grows (fig.1b, curves 4 and 5).

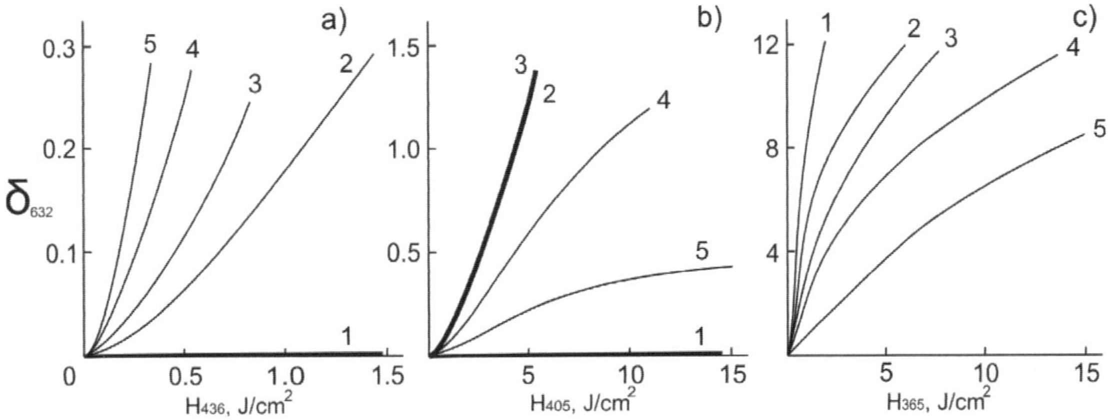

Fig.1. The initial kinetics sections of photo-induced phase retardation δ_{632} within PVMCi layer with BNT addition with concentration of 1–0%; 2–0.5%; 3–1.0%; 4–2.0%; 5–4.0% as energy function H_{exp} with radiation λ_{exp} – (a) 436 nm, (b) 405 nm и (c) 365 nm.

An unsensitized PVCi layer (fig.2, curve 1), as well as the similar PVMCi level, doesn't absorb with the wave-length more than 400 nm and, as the result, doesn't create POA effect under the radiation of λ_{exp}=405 nm. It appears only with sensitizer, and moreover, the speed and the value of the induced phase retardation increases due to the increased Michler's kethon (MK) sensitizer concentration in the layer.

In PVMCi film without the sensitizer and with the radiation of λ_{exp}>400 nm (curve 1 in figure 1a and 1b the POA effect is missing, i.e. each reaction component doesn't create any POA effect by itself. PVCi doesn't absorb with the wave-length more than 405 nm, and MK sensitizer's molecules are photochemically stable.

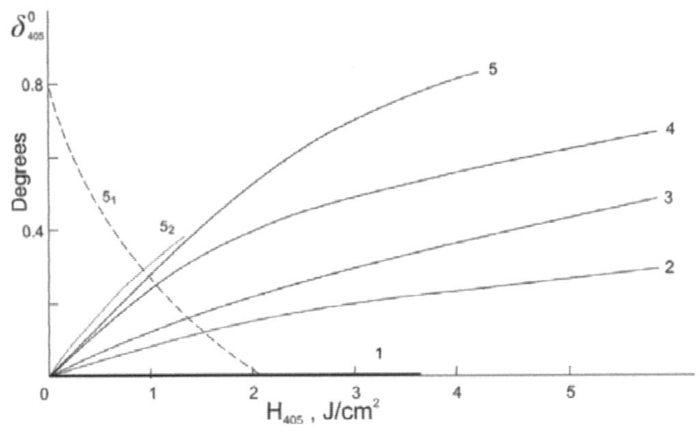

Fig.2. Photo-induced phase retardation kinetics δ_{405} in PVCi layer with MK addition with the concentration of 1–0%; 2–0.5%; 3–1.0%; 4–2.0%; 5–4.0% as energy function with radiation of λ_{exp} = 405 nm.

The induced anisotropy can be optically "erased" as shown in figure 2 curve 5_1 and restored as in fig.2 curve 5_2.

Acknowledgements
This research was supported by RFBR, Research Grant No. 17–47–500757, and Ministry of Education, Investments and Innovations in Moscow Region, Russian Federation Presidential Grant No. MK–7359.2016.9.

IOP Conf. Series: Journal of Physics: Conf. Series **867** (2017) 012039 doi:10.1088/1742-6596/867/1/012039

References

[1] SU patent №SU 1 769 607 A1 Polarization Sensitive Photographic material Kozenkov, V.M., Kisilitsa, P.P., Ganuschak, N.I., Katyshev, E.G., Naumova, N.A., Orlova, T.N., Obushak, N.D., Shulev, Y.V. 1995.
[2] Kozenkov, V.M., Katyshev, E.G., Barachevsky, V.A., Kisilitsa, P.P., Naumova, N.A. Scientific and applied photography and cinematography magazine. 1985. SU 4. p.281
[3] Kozenkov, V.M., Chigrinov, V.G., Kwok, H. S. Mol. Cryst. Liq. Cryst. 2004. V.409 p.251.
[4] Kozenkov, V.M., Spakhov, A.A., Belyaev, V.V., Chausov D.N. SID Digest 2016. V.47, p.1645.
[5] Belyaev, V.V., Solomatin, A.S., Chausov D.N. Opt. Exp. 2013. V.21 p.4244.
[6] Furumi S., Ichimura K. Phys. Chem. Chem. Phys. 2011. V.13. p.4919.

IOP Conf. Series: Journal of Physics: Conf. Series **867** (2017) 012040 doi:10.1088/1742-6596/867/1/012040

Characteristics of Einstein oscillator, electron-phonon interaction, and band-edge absorption in $LiNbO_3$:In

Feifei Xin

College of Physics and Materials Science, Tianjin Normal University, Tianjin, 300387, China

E-mail: `cindyfeifei@163.com`

Abstract. Based on the Einstein oscillator model and the Urbach rule, we study the electron-phonon interaction and the band-edge structure of highly doped $LiNbO_3$:In. Significant reduction of the Einstein oscillator energy and the electron-phonon interaction strength appears in highly doped $LiNbO_3$:In compared with the undoped one. The absorption band near to the UV band edge in highly doped $LiNbO_3$:In, which is much stronger than that in the undoped $LiNbO_3$ crystal, may be related to the defect clusters (O^{2-}-V_{Li}^{-}) and (O^{2-}-In_{Nb}^{2-}) and the micromechanism of the UV photorefractivity in $LiNbO_3$:In.

It was found that in lithium niobate ($LiNbO_3$, LN) crystal the optical damage in the visible is significantly suppressed by highly In-doping, while the photorefractivity in highly doped LN:In is greatly enhanced in the UV [1, 2]. In this paper, we report on the thermal induced spectra properties of the fundamental band edge and the deep level defect structures of highly doped LN:In.

Congruent LN crystals with In-doping concentrations of 0.0 and 5.0 mol% in melt was prepared and labeled as CLN and CIn5, respectively. Both of the crystals are cut in the form of Y-oriented plates with a thickness of 0.5 mm. The UV band edge absorption spectra in the temperature range from 3.8 K to 973 K were measured for our samples. The band energy gap E_g, defined as the photon energy at which $\alpha = 70\ cm^{-1}$, decreases with the increase of temperature, which can be well described by the Bose-Einstein expression $E_g(T) = E_{gBE}(0) - (2a_B)/[exp(E_{pBE}/k_BT) - 1]$, as shown in Fig. 1(a) [3]. The fundamental band gap $E_{gBE}(0)$ has a significant increase in CIn5 and the Einstein oscillator energy E_{pBE} drops abruptly in CIn5 as compared with CLN[4]. The electron-phonon interaction strength weakens dramatically in the CIn5, which would theoretically lead to an increase in the photoconductivity. Note that the photoconductivities of highly doped LN:In are already found to be remarkably enhanced in both visible and UV region [1, 2], this result, therefore, gives a fundamental microscopic picture for the observed optical properties in highly doped LN:In crystals.

The absorption edge tails of both samples at temperatures higher than 400 K are consistent with the widely observed Urbach rule [5]. However, as the temperature goes down below $\sim$ 400 K and the absorption edge shifts towards higher energy, an additional absorption band shows up at the bottom of the absorption edge tail. Typical absorption spectra at a cryogenic temperature of 3.8 K for both samples are shown in Fig. 1(b). It is the appearance of this new absorption band that deforms the Urbach tail at this spectral region. In order to show the true spectral structure of this absorption band, we subtract the Urbach tail from the measured spectra and get $\Delta\alpha = \alpha - \alpha_0 \exp\left[\frac{\sigma}{k_BT^*}(\hbar\omega - \hbar\omega_0)\right]$, where α_0, σ, ω_0 and T^* are calculated

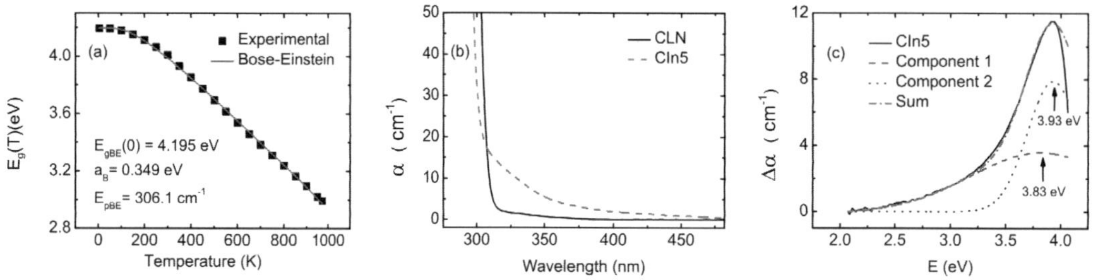

Figure 1. (a) The measured temperature dependence of the band energy gap of CIn5 with the corresponding fitting curves by the Bose-Einstein expression. (b) The absorption spectra of CLN and CIn5 crystals at 3.8 K. (c) The spectral profile of $\Delta\alpha$ measured at 3.8 K for CIn5 with theoretical fit of Gaussian profiles peaked at 3.83 eV and 3.93 eV, respectively

by fitting the remaining spectrum without deformation to the Urbach rule. The the calculated spectral profile of $\Delta\alpha$ measured at 3.8 K for CIn5 is shown in Fig. 1(c). We find that the peak of the absorption band $\Delta\alpha$ of CIn5 is higher than that of CLN [4], which has the same tendency as the enhancement of the UV photorefractivity in highly doped LN:In [2]. Therefore, the corresponding defect centers should be responsible for the photorefraction and the light-induced charge transport process in the UV in highly doped LN:In. Moreover, for CIn5, the absorption band $\Delta\alpha$ at 3.8 K can be decomposed into two Gaussian components peaked at 3.83 eV and 3.93 eV as shown in Figs. 1 (c), while typical result for CLN can be well fitted by a Gaussian spectral profile with its peak energy at 3.83 eV [4]. Just as the existence of defect cluster (O^{2-}-Mg_{Nb}^{3-}) in LN:Mg [4], it is reasonable to suppose that, besides (O^{2-}-V_{Li}^{-}) defect cluster, (O^{2-}-In_{Nb}^{2-}) will also form deep energy levels close to the valence band in the band gap of LN:In. It was found that the two deep energy levels in highly doped LN:Mg are centered at 3.83 eV and 4.03 eV, while in CIn5, one can fit the absorption band of $\Delta\alpha$ by the sum of two Gaussian components centered at 3.83 eV and 3.93 eV. Since In_{Nb}^{2-} has a valence magnitude between V_{Li}^{-} and Mg_{Nb}^{3-}, with respect to the crystal lattice, the two energy levels in highly doped LN:In crystal are closer to each other than those in LN:Mg.

In summary, the UV band edge spectra and the UV photo-sensitive defect centers of highly doped LN:In has been studied. The Einstein oscillator energy and the electron-phonon interaction strength in highly doped LN:In decreased obviously compared with CLN, which could be regarded as the microscopic origin of the significant increase in photoconductivity in highly doped LN:In. A new absorption band at the bottom of the Urbach tail appears in CIn5 at low temperatures, and it is attributed to (O^{2-}-V_{Li}^{-}) and (O^{2-}-In_{Nb}^{2-}) defect clusters, which may determine the band-edge optical properties of highly doped LN:In..

Acknowledgments

This work is supported by the Doctoral Foundation of Tianjin Normal University (Grant No.135202XB1607).

References

[1] T. Volk and M. Wöhlecke 2008 *Lithium Niobate: Defects, Photorefraction and Ferroelectric Switching* (Berlin: Springer-Verlag)
[2] F. Xin, G. Zhang, F. Bo, H. Sun, Y. Kong, J. Xu, T. Volk, and N. Rubinina 2010 *J. Appl. Phys.* **107** 033113
[3] L. Viña, S. Logothetidis, and M. Cardona 1984 *Phys. Rev.* B **30** 1979
[4] F. Xin, Z. Zhai, X. Wang, Y. Kong, J. Xu, and G. Zhang 2012 *Phys. Rev.* B **86** 165132
[5] F. Moser and F. Urbach 1956 *Phys. Rev.* **102** 1519